HOME HEATING & COOLING

Other Publications:

THE TIME-LIFE GARDENER'S GUIDE

MYSTERIES OF THE UNKNOWN

TIME FRAME

FIX IT YOURSELF

FITNESS, HEALTH & NUTRITION

SUCCESSFUL PARENTING

HEALTHY HOME COOKING

UNDERSTANDING COMPUTERS

LIBRARY OF NATIONS

THE ENCHANTED WORLD

THE KODAK LIBRARY OF CREATIVE PHOTOGRAPHY

GREAT MEALS IN MINUTES

THE CIVIL WAR

PLANET EARTH

COLLECTOR'S LIBRARY OF THE CIVIL WAR

THE EPIC OF FLIGHT

THE GOOD COOK

WORLD WAR II

HOME REPAIR AND IMPROVEMENT

THE OLD WEST

HOME HEATING & COOLING

TIME-LIFE BOOKS
ALEXANDRIA, VIRGINIA

Fix It Yourself was produced by
ST. REMY PRESS

MANAGING EDITOR	Kenneth Winchester
MANAGING ART DIRECTOR	Pierre Léveillé

Staff for *Home Heating & Cooling*

Series Editor	Kathleen M. Kiely
Editor	Dianne Thomas
Series Art Director	Diane Denoncourt
Art Director	Odette Sévigny
Research Editor	Michael Mouland
Designers	Maryse Doray, Solange Pelland
Editorial Assistant	Cathleen Farrell
Contributing Writers	Margaret Caldbick, Kent Farrell, James Fehr, Harriett Fels, Emer Killean, Grant Loewen, Daniel McBain
Electronic Designer	Daniel Bazinet
Contributing Illustrators	Gérard Mariscalchi, Jacques Proulx
Technical Illustrators	Nicolas Moumouris, Robert Paquet
Cover	Robert Monté
Index	Christine M. Jacobs
Administrator	Denise Rainville
Coordinator	Michelle Turbide
Systems Manager	Shirley Grynspan
Systems Analyst	Simon Lapierre
Studio Director	Maryo Proulx

Time-Life Books Inc. is a wholly owned subsidiary of
TIME INCORPORATED

FOUNDER	Henry R. Luce 1898-1967
Editor-in-Chief	Jason McManus
Chairman and Chief Executive Officer	J. Richard Munro
President and Chief Operating Officer	N. J. Nicholas Jr.
Editorial Director	Ray Cave
Executive Vice President, Books	Kelso F. Sutton
Vice President, Books	George Artandi

TIME-LIFE BOOKS INC.

EDITOR	George Constable
Executive Editor	Ellen Phillips
Director of Design	Louis Klein
Director of Editorial Resources	Phyllis K. Wise
Editorial Board	Russell B. Adams Jr., Dale M. Brown, Roberta Conlan, Thomas H. Flaherty, Lee Hassig, Donia Ann Steele, Rosalind Stubenberg, Henry Woodhead
Director of Photography and Research	John Conrad Weiser
Asst. Director of Editorial Resources	Elise Ritter Gibson
PRESIDENT	Christopher T. Linen
Chief Operating Officer	John M. Fahey Jr.
Senior Vice Presidents	Robert M. DeSena, James L. Mercer
Vice Presidents	Stephen L. Bair, Ralph J. Cuomo, Neal Goff, Stephen L. Goldstein, Juanita T. James, Hallett Johnson III, Carol Kaplan, Susan J. Maruyama, Robert H. Smith, Joseph J. Ward
Director of Production Services	Robert J. Passantino

Editorial Operations

Copy Chief	Diane Ullius
Production	Celia Beattie
Library	Louise D. Forstall
Correspondents	Elizabeth Kraemer-Singh (Bonn); Maria Vincenza Aloisi (Paris); Ann Natanson (Rome).

THE CONSULTANTS

Consulting Editor **David L. Harrison** served as an editor for several Time-Life Books do-it-yourself series, including *Home Repair and Improvement*, *The Encyclopedia of Gardening* and *The Art of Sewing*.

Evan Powell is Director of Chestnut Mountain Research Inc in Taylors, South Carolina, a firm that specializes in the development and evaluation of heating equipment as well as home and building products. He is a contributing editor to several do-it-yourself magazines, and the author of a book on heating and cooling.

Elliot Levine, special consultant for Canada, is a mechanical engineer and third-generation plumbing and heating specialist. He and his brother Neal operate Levine Brothers Plumbing in Montreal.

Eldon Wilson, a refrigeration specialist for more than 30 years, is the president of Electro Aide Inc. in St. Laurent, Quebec, a firm that specializes in the installation and servicing of heat pumps and air conditioning systems for residential, commercial and industrial use.

Library of Congress Cataloging-in-Publication Data
Home heating & cooling
      p. cm. – (Fix it yourself)
  Includes index.
  ISBN 0-8094-6244-3
  ISBN 0-8094-6245-1 (lib. bdg.)
1. Dwellings—Heating and ventilation—Maintenance and repair—Amateurs' manuals. 2. Dwellings—Air conditioning—Maintenance and repair—Amateurs' manuals. 3. Heating—Maintenance and repair—Amateurs' manuals. I. Time-Life Books. II. Title: Home Heating & Cooling. III. Series.
TH7225.H59   1988
697—dc19   88-14121
                       CIP

For information about any Time-Life book, please write:
Reader Information
Time-Life Customer Service
P.O. Box C-32068
Richmond, Virginia
23261-2068

CONTENTS

HOW TO USE THIS BOOK

Home Heating and Cooling is divided into three sections. The Emergency Guide on pages 8-13 provides information that can be indispensable, even lifesaving, in the event of a household emergency. Take the time to study this section *before* you need the important advice it contains.

The Repairs section — the heart of the book — is a comprehensive system for troubleshooting and repairing heating and cooling units. Pictured below are four sample pages from the chapter on oil burners, with captions describing the various features of the book and how they work. If your oil-fired heating system is not producing sufficient heat, for example, the Troubleshooting Guide will offer a number of possible causes. The problem may be a dirty or faulty photocell flame detector. In that case, you will be directed to page 55 for detailed, step-by-step directions for removing, cleaning and replacing the photocell.

Each job has been rated by degree of difficulty and the average time it will take for a do-it-yourselfer to complete. Keep in mind that this rating is only a suggestion. Before deciding whether you should attempt a repair, first read all the

---

Introductory text
Describes principles of operation, most common breakdowns and basic safety precautions. Suggests other chapters that should be consulted.

Exploded and cutaway diagrams
Locate and describe the various components of the unit. Variations are pictured where applicable.

Troubleshooting Guide
To use this chart, locate the symptom that most closely resembles your system problem, review the possible causes in column 2, then follow the recommended procedures in column 3. Simple fixes may be explained on the chart; in most cases you will be directed to an illustrated, step-by-step repair sequence.

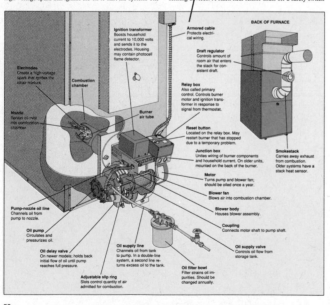

Degree of difficulty and time
Rate the complexity of each repair, and how much time the job should take for a homeowner with average do-it-yourself skills.

Special tool required
Some repairs require a multitester or other diagnostic tool.

instructions carefully. Then be guided by your own confidence, and the tools, time and replacement parts available to you. For more complex or time-consuming repairs, such as replacing the motor, you may wish to call for professional service. You will still have saved time and money by diagnosing the problem yourself.

Most repairs in *Home Heating and Cooling* can be made with simple tools including screwdrivers, wrenches, pliers and a multitester. Any special tool required is indicated in the Troubleshooting Guide. Basic tools — and the proper way to use them — are presented in Tools & Techniques *(page 132)*. If you are a novice at home repair, read this section and the chapter called Heating and Cooling Systems *(page 14)* in preparation for a job.

Repairing a heating or cooling unit is easy and safe if you work logically and follow the precautions in each chapter. Before beginning most repairs, turn off power to the unit; on a gas burner, turn off the gas. Where recommended, discharge all capacitors in the unit. Store fasteners and small parts in labeled containers, and label wires before disconnecting them.

Name of repair
You will be referred by the Troubleshooting Guide to the first page of a specific repair job.

Step-by-step procedures
Follow the numbered repair sequence carefully. Depending on the result of each step, you may be directed to a later step, or to another part of the book, to complete the repair.

Lead-ins
Bold lead-ins summarize each step or highlight the key action pictured in the illustration.

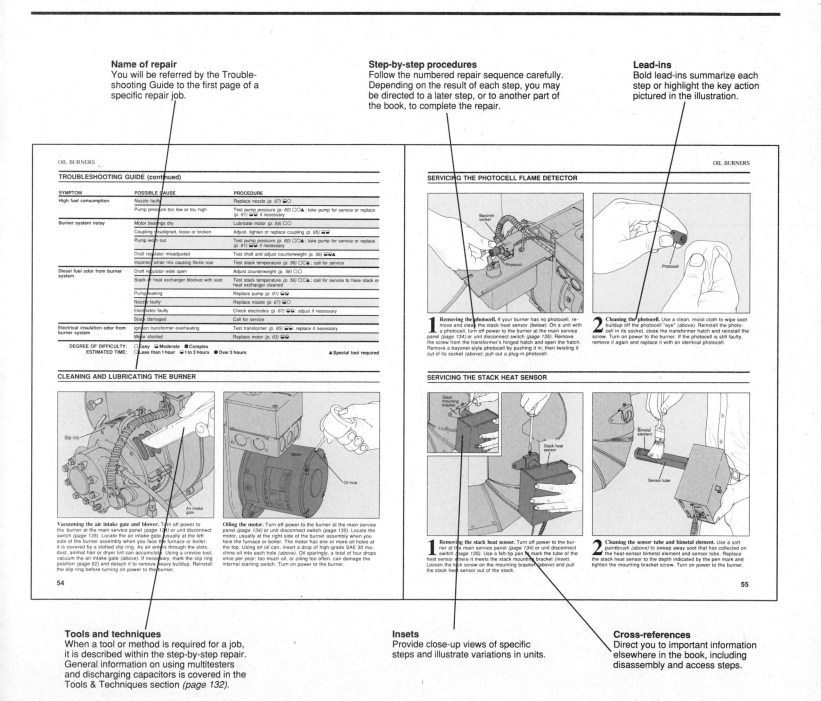

Tools and techniques
When a tool or method is required for a job, it is described within the step-by-step repair. General information on using multitesters and discharging capacitors is covered in the Tools & Techniques section *(page 132)*.

Insets
Provide close-up views of specific steps and illustrate variations in units.

Cross-references
Direct you to important information elsewhere in the book, including disassembly and access steps.

EMERGENCY GUIDE

Preventing heating and cooling emergencies. Heating and cooling systems today are so reliable that homeowners take safe, uninterrupted service for granted — until an emergency occurs. If properly installed, your system will rarely pose a hazard, as long as it is also properly maintained. The routine maintenance recommended in this book can extend the life of your system, help prevent costly repairs and, most important, lower the risk of heating failure in the middle of winter. Also consult the owner's manual for your system, to learn the manufacturer's recommended maintenance schedules.

Before working, turn off power to the unit *(page 134)* whenever recommended; some repairs also require testing to make sure power is off. Alert family members by affixing a tag labeled "DO NOT TURN ON" to the main service panel and the unit disconnect switch. Do not attempt to shut off power if the work area is flooded; call a professional. When working on a gas or oil burner, turn off the fuel supply where directed.

The pressurized liquid refrigerant in heat pump and air conditioning coils can seriously frostbite exposed skin. When working near refrigerant lines or coils, wear safety goggles and work gloves. Use tools carefully; a tool that slips could rupture the refrigerant tubing.

The Troubleshooting Guide on page 9 places emergency procedures at your fingertips. Some involve routine techniques that you and your family should know; others are stopgap fixes, just until you can get parts or professional service. Read the Guide before you need it; also read the list of safety tips at right. Familiarize yourself with Tools & Techniques *(page 132)*, which describes the safe use of tools.

When in doubt about your ability to handle an emergency, don't hesitate to call for help. Post the telephone numbers for the fire department, gas or oil supplier and electric company near the telephone.

NATURAL GAS IN THE HOME

A well-maintained gas furnace is one of the safest systems for heating your home, but it is important to know its hazards. In high enough concentrations, natural gas is explosive; a gas leak in the home can also lead to asphyxiation and cause natural-gas poisoning.

For consumer safety, gas companies are required by law to add mercaptans — sulphurous chemicals — to natural gas; the odor warns homeowners of a leak. If the gas odor lingers, do not relight a pilot or strike any matches. Avoid turning on lights; switches can spark. Open doors and windows to ventilate the room, and turn off the gas supply to the burner until the odor dissipates. If the odor persists, leave the house and call the gas company for service.

If you return home and smell gas, DO NOT enter the building; call the gas company from a neighbor's house. Gas inhalation symptoms include nausea and choking; take victims outdoors immediately, and call for medical help.

SAFETY TIPS

1. Before attempting any repair in this book, read the entire procedure. Familiarize yourself with the specific safety information presented in each chapter.

2. Use the right tools for the job, including those with insulated handles, and do not substitute tools. Wear the proper protective gear for the job. Refer to Tools & Techniques *(page 132)*.

3. Post the telephone numbers of emergency medical help, the fire department and all utility companies near the telephone.

4. Familiarize yourself and family members with the correct procedures for shutting off your home's power, water and fuel supplies. For quick identification, label the main gas and water shutoff valves.

5. Label your main service panel with the locations of the heating and cooling system circuit breakers or fuses *(page 134)*.

6. Wherever instructed, turn off electrical power to a unit under repair *(page 134)*. Affix a warning sign on unit controls and power cords, and the main service panel, to remind others that you are working.

7. Discharge capacitors safely *(page 140)* before working in the electrical service box of an air conditioning or heat pump unit.

8. Don't work in wet or flooded conditions. If the ground or floor is damp, stand on dry boards and wear an insulated glove to shut off power at the main service panel *(page 134)*.

9. If the odor of gas does not dissipate, do not attempt to relight a gas-burner pilot light; do not light matches or touch electrical outlets or switches. Ventilate the area, leave the house, and call the gas company. Never enter a house that is filled with the odor of gas.

10. Use only replacement parts and wiring of the same specifications as the original. Look for the UL (Underwriters' Laboratories) label on new parts. If in doubt, consult the manufacturer, or take the original part to the dealer.

11. The diagnosis and repair of some heating and cooling problems are beyond the scope of the homeowner. Never attempt a repair that you are unqualified to do; call for professional service where recommended.

12. Avoid damage to refrigerant coils and lines; have them serviced only by a professional.

13. Install smoke detectors and fire extinguishers in your home.

14. Catch draining fuel or water in a sturdy basin, and wipe up spills immediately. Discard used oil and other flammable substances according to local environmental regulations.

15. When reassembling a unit, make sure that all wiring connections are tight, that no wires are pinched and that all fuel lines are secure.

16. Keep furniture and curtains away from baseboard heaters and make sure that air distribution registers are unblocked.

17. Ask the fuel company for an annual system checkup, including a measurement of the carbon monoxide level in your home. If the safety of your heating or cooling system or of any repair is in doubt, call for professional advice.

18. Keep the work area clean and well lit. Keep children away and do not allow them to touch the controls of heating and cooling systems.

19. Let all heating and cooling appliances cool down before starting repairs.

TROUBLESHOOTING GUIDE

SYMPTOM	PROCEDURE
Power failure	Turn off all appliances with motors or heating elements (furnace, air conditioner, baseboard heater) to prevent overloading system when power is restored
	Check service panel; replace blown fuses or reset tripped circuit breakers *(p. 134)*
	Have emergency supplies on hand, including a small space heater and a lantern, flashlight or candles. A portable generator can provide a limited amount of emergency power *(p. 13)*
Electrical shock	If victim is breathing and has a pulse, begin cardiopulmonary resuscitation (CPR) if you are qualified to do so. Otherwise, place victim in recovery position *(p. 10)* and call for help
Unit sparking or hot to the touch	Turn off power at main service panel *(p. 9)*; or unplug cord, protecting your hand *(p. 10)*
Refrigerant burn	Apply non-adhesive sterile gauze *(p. 12)*; seek medical help immediately
Heating or cooling unit on fire	Use ABC or BC fire extinguisher *(p. 11)*; call fire department
Gas odor in house	Turn off gas at main shutoff valve *(p. 10)* and ventilate room; do not touch electrical outlets or switches
	If odor persists, evacuate house and call gas company
No heat due to clogged air filter in air distribution system	Check air filter *(Air Distribution, p. 24)*. If new air filter is not available, clean filter *(p. 12)*; replace filter as soon as possible
No heat due to faulty thermostat	Troubleshoot thermostat *(System Controls, p. 16)*; if faulty, and new thermostat is not available, jumper thermostat *(p. 12)*; replace thermostat as soon as possible
No heat due to faulty aquastat in water distribution system	Test aquastats *(Water Distribution, p. 39)*; if faulty, and new aquastat is not available, jumper aquastat *(p. 12)*. **Caution:** If burner aquastat is jumpered, allow system to run for only one 30-minute cycle every hour. Replace aquastat as soon as possible
	Call for service

TURNING OFF POWER TO THE HOUSE

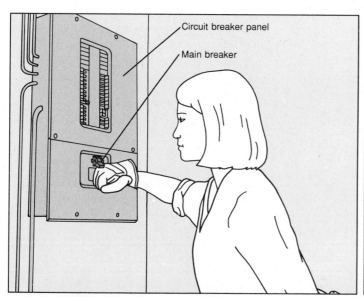
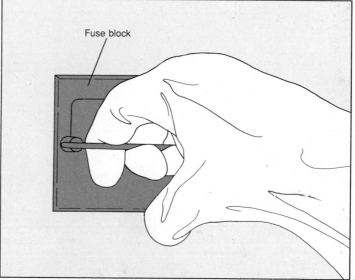

Disconnecting power at the main service panel. If the floor below the service panel is wet, stand on a dry board or rubber mat, or wear rubber boots. Use a wooden broom handle or wear heavy work gloves; work with one hand only to protect your body from becoming a path for electrical current, and keep the other hand in your pocket or behind your back. At a circuit breaker panel, flip off the main breaker *(above, left)*. As an added precaution, use your knuckle; any shock will jerk your hand away from the panel. At a fuse panel, remove the main fuse block by gripping its metal handle and pulling it from the box *(above, right)*. On a panel with more than one fuse block, remove them all. Some fuse panels have a shutoff lever instead; pull down the lever to turn off power to the house.

ELECTRICAL EMERGENCIES

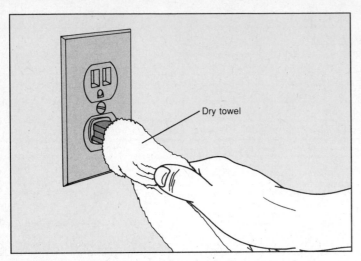

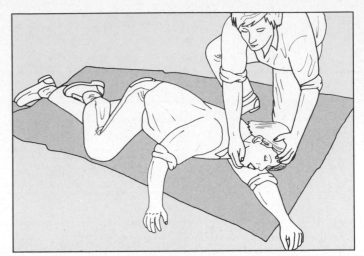

Pulling the power cord. Caution: If the floor is wet, or the outlet itself is sparking or burning, do not touch the cord or unit; turn off power to the circuit at the service panel *(page 9)*. If the unit sparks, shocks you, feels hot or is burning, disconnect the plug. Protect your hand with a thick, dry towel or a heavy work glove. Without touching the outlet, grasp the cord with one hand several inches from the plug *(above)* and pull it out. Repair the problem before plugging in the unit.

Handling a victim of electrical shock. A person who contacts live current may be thrown back from the source. But muscles may contract around a wire or appliance. If the victim is stuck, do not touch him; quickly pull the plug or shut off power *(page 9)*, or use a wooden broomstick or chair to separate him from the circuit. Call for help immediately. If an unconscious victim is breathing and has not received back injuries, place him in the recovery position *(above)*. Tilt the head back with the face to one side and tongue forward to maintain an open airway.

GAS EMERGENCIES

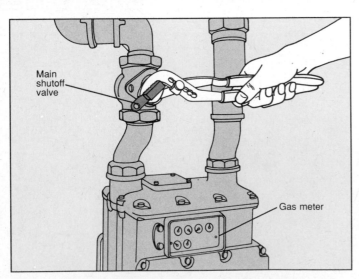

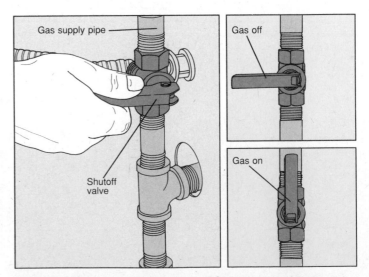

Turning off the main gas supply. The main shutoff valve is located on a pipe at the utility meter. Use an adjustable wrench to turn it off *(above)*; when the handle is perpendicular to the pipe, the gas is off. Ventilate the area. Do not use electrical switches—a spark could ignite the gas. If the gas does not dissipate, evacuate the house, then call the gas company.

Shutting off the gas at the boiler or furnace. Turn off the unit shutoff valve by gripping the valve lever and turning it *(above)* until it is perpendicular to the gas supply pipe leading to the unit *(inset)*. If the shutoff valve is a surface-mounted control, switch it to the OFF position. Tag the gas shutoff valve for easy identification in an emergency.

WATER EMERGENCIES

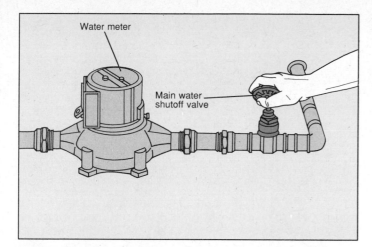

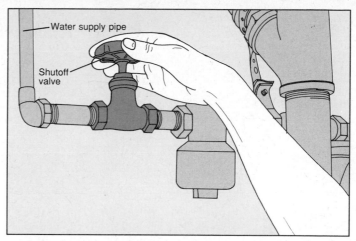

Shutting off the main water valve. If water is leaking through the wall or ceiling from an undetermined source, turn off the house water supply at the main shutoff valve *(above)*. It is usually located near the water meter or where the main water-supply pipe enters the house. **Caution:** Never enter a flooded basement to turn off water or power; call the utility company.

Shutting off water at the boiler. Locate the shutoff valve on the water pipe that supplies the boiler. Grip the valve handle and turn it clockwise to prevent water from flowing to the boiler *(above)*. Tag the valve for easy identification in an emergency.

FIRE EMERGENCIES

Extinguishing a fire. Have someone call the fire department immediately. If flames or smoke are coming from the walls or ceiling, leave the house to call for help. To snuff a small electrical fire, use a dry-chemical fire extinguisher rated ABC or BC. Stand near an exit, 6 to 10 feet from the fire. Holding the extinguisher upright, pull the lock pin out of the handle and aim the nozzle at the base of the flames. Squeeze the handle and spray in a quick side-to-side motion *(left)* until the fire is completely out. Watch for "flashback," or rekindling, and be prepared to spray again. You may also have to turn off power at the service panel *(page 9)* to remove the source of heat at the fire. If the fire spreads, leave the house.

TREATING A LIQUID REFRIGERANT BURN

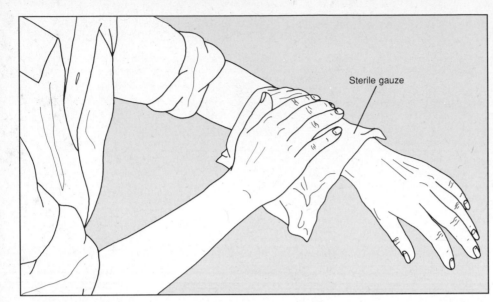

Sterile gauze

Applying a lukewarm compress. A liquid refrigerant "burn" is actually a form of severe frostbite, caused when the skin comes in contact with an extremely cold fluorocarbon chemical under high pressure. If the affected area is red, indicating a first-degree burn, but shows no sign of blistering, apply a layer of nonadherent, sterile gauze moistened with lukewarm water. **Caution:** Do not apply antiseptic sprays, ointments or chemical neutralizers of any kind. A minor wound will heal itself. If blisters appear, indicating a second-degree burn, keep the area dry. Protect the burn with a layer of nonadherent sterile gauze and seek medical help immediately.

STOPGAP PROCEDURES FOR RESTORING HEAT

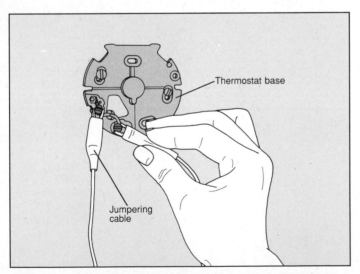

Thermostat base

Jumpering cable

Jumpering a low-voltage thermostat. If the cause of a heating system failure is a faulty thermostat, you can jumper the thermostat, bypassing it to get the system going temporarily. Turn off power to the heating system at the main service panel *(page 134)* and unit disconnect switch *(page 135)*. Access the low-voltage connections on the thermostat base *(page 16)*. Locate the terminals marked "R" and "W" connected to the red and white wires from the wall. Attach one clip of a jumpering cable *(page 132)* to each terminal. Turn on power to the system. When the house temperature has reached a comfortable point, turn off power. Repeat as necessary. Replace the thermostat as soon as possible *(page 22)*.

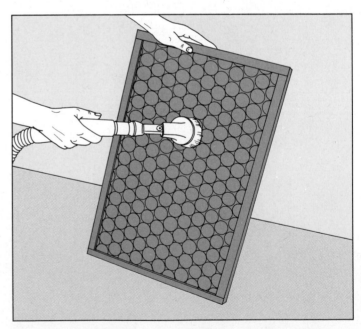

Cleaning a disposable air filter. If an air distribution system malfunctions due to a clogged filter, and you do not have a spare filter on hand, take out the filter *(page 26)* and use a vacuum cleaner with a brush attachment to remove debris embedded in the filter's fibers *(above)*. Reinstall the filter; replace it with a new one as soon as possible *(page 26)*.

STOPGAP PROCEDURES FOR RESTORING HEAT (continued)

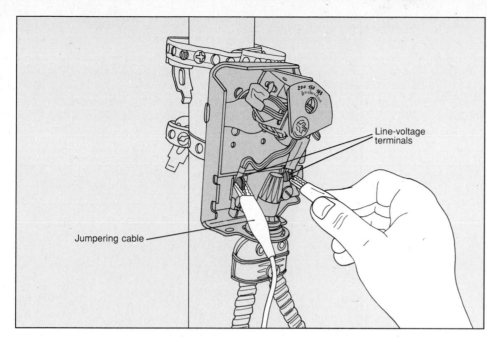

Line-voltage terminals

Jumpering cable

Jumpering an aquastat. If you have diagnosed a faulty burner aquastat or pump aquastat *(Water Distribution, page 39)*, you can get the system operating temporarily — just until a replacement part is available. Turn off power to the boiler at the main service panel *(page 134)* and unit disconnect switch *(page 135)*. Remove the cover on the aquastat to access the line-voltage terminal screws *(page 47)*. Use a voltage tester to confirm that power is off *(page 48)*. Attach one clip of a jumpering cable of the proper gauge *(page 132)* to each line-voltage terminal on the aquastat *(left)*. Turn on power to the system. Do not touch the terminals or jumpering cable while power is on. **Caution:** When the burner aquastat is jumpered, the safety shutoff sytem is bypassed; allow the system to run for only 30 minutes every hour. Turn off power to the system immediately if the pressure gauge on the boiler exceeds 30 psi (pounds per square inch).

COPING WITH HEAT FAILURE

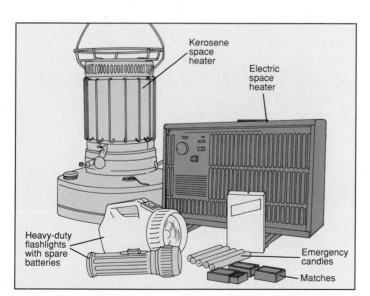

Kerosene space heater

Electric space heater

Heavy-duty flashlights with spare batteries

Emergency candles

Matches

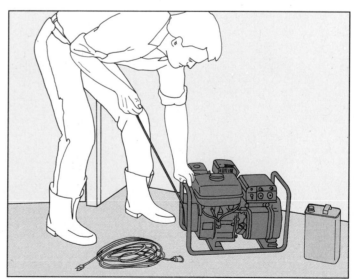

Emergency supplies. Plan in advance for power failure or heating malfunction. Candles, a reliable source of light, provide a surprising amount of warmth. Keep candles, matches and a flashlight in a familiar location. A kerosene heater rated for indoor use is a safe temporary source of heat, provided the manufacturer's instructions are strictly followed. Be sure to keep fuel on hand. **Caution:** When using a fuel-fired heater, always open a window slightly for ventilation. A portable electric heater is safest, if power is available.

A portable electrical generator. A gasoline generator can be rented to deal with long-term power failure. A generator must be operated in a dry, sheltered, ventilated area; a garage or porch is ideal. Only appliances that can use extension cords should be attached to a generator. Permanent hookup to the house wiring must be left to a professional. A 2200-watt unit *(above)* is large enough to operate a refrigerator or electric space heater and several lamps. To run appliances that have a compressor, a generator with surge capacity is needed.

HEATING AND COOLING SYSTEMS

No matter what heating or cooling system you have, your home's climate is unique. Your comfort depends on the house's insulation, location and exposure, as well as how your heating and cooling system is maintained.

The system's design may be as simple as a single baseboard heater in each room, or more complicated, such as an air distribution system using a heat pump plus an oil-fired furnace and central humidifier. All systems, however, have four things in common. Each system has a *heat producer*—an oil or gas burner or an electric heating element, and a *heat exchanger*—a furnace where air is heated or a boiler where water is heated. A system's *heat distributor* may be the ducts and registers that circulate air throughout the home, or the pipes and radiators that circulate water. Finally, each system has a *control*—a thermostat or humidistat. When troubleshooting a heating or cooling problem, make sure to refer to all chapters pertaining to your system.

Listen to how your system sounds when it operates properly. This will give you important clues when something goes wrong: Does the blower start and stop too frequently? Do you

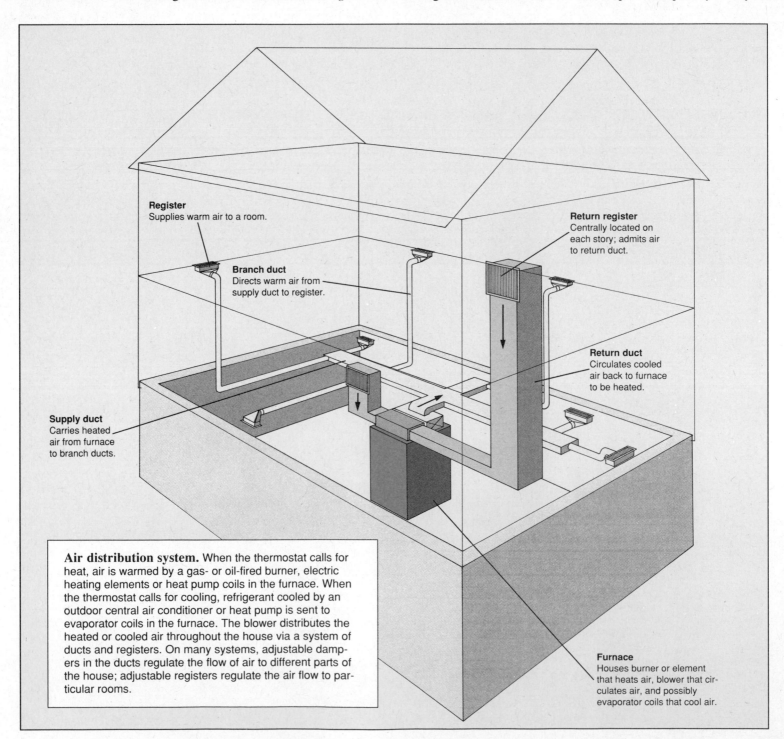

Register
Supplies warm air to a room.

Branch duct
Directs warm air from supply duct to register.

Return register
Centrally located on each story; admits air to return duct.

Return duct
Circulates cooled air back to furnace to be heated.

Supply duct
Carries heated air from furnace to branch ducts.

Furnace
Houses burner or element that heats air, blower that circulates air, and possibly evaporator coils that cool air.

Air distribution system. When the thermostat calls for heat, air is warmed by a gas- or oil-fired burner, electric heating elements or heat pump coils in the furnace. When the thermostat calls for cooling, refrigerant cooled by an outdoor central air conditioner or heat pump is sent to evaporator coils in the furnace. The blower distributes the heated or cooled air throughout the house via a system of ducts and registers. On many systems, adjustable dampers in the ducts regulate the flow of air to different parts of the house; adjustable registers regulate the air flow to particular rooms.

hear any unusual noises? The answers to questions like these can help you locate a problem more easily.

Pictured below are two representative home heating and cooling systems. Remember that these systems can be assembled and wired in a variety of ways, depending on the model, its age, and the space it occupies. If necessary, ask a service person to help you identify your system's components.

For many families, heating and cooling is a major monthly expense. By keeping your system well maintained, you can minimize the energy it uses. To use that energy most efficient-ly, make sure your home is well insulated, and weather-strip, caulk, or add storm windows and doors if necessary. Lined draperies help insulate windows and stop drafts; keep draperies open during the day in winter to let the sunshine in, and close them at night. Shut off most of the heat to unused rooms by closing registers, dampers or radiators. Set your thermostat back at night and when you're not at home, or install an electronically timed thermostat. Do not allow furniture or draperies to block registers, radiators, convectors or baseboard heaters. Keep the fireplace damper closed when not in use.

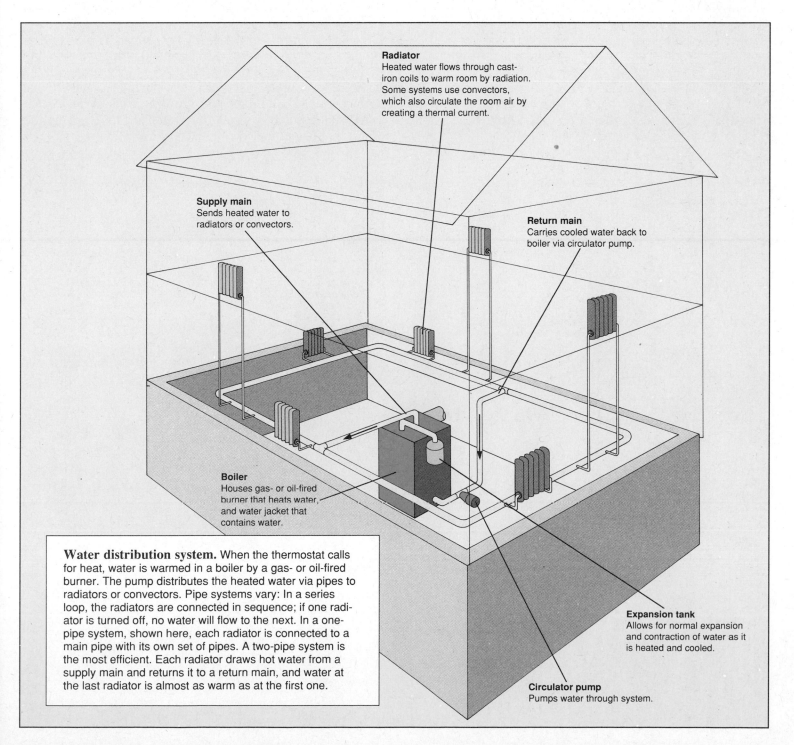

Radiator
Heated water flows through cast-iron coils to warm room by radiation. Some systems use convectors, which also circulate the room air by creating a thermal current.

Supply main
Sends heated water to radiators or convectors.

Return main
Carries cooled water back to boiler via circulator pump.

Boiler
Houses gas- or oil-fired burner that heats water, and water jacket that contains water.

Expansion tank
Allows for normal expansion and contraction of water as it is heated and cooled.

Circulator pump
Pumps water through system.

Water distribution system. When the thermostat calls for heat, water is warmed in a boiler by a gas- or oil-fired burner. The pump distributes the heated water via pipes to radiators or convectors. Pipe systems vary: In a series loop, the radiators are connected in sequence; if one radiator is turned off, no water will flow to the next. In a one-pipe system, shown here, each radiator is connected to a main pipe with its own set of pipes. A two-pipe system is the most efficient. Each radiator draws hot water from a supply main and returns it to a return main, and water at the last radiator is almost as warm as at the first one.

SYSTEM CONTROLS

The thermostat and humidistat provide fingertip control of complicated systems. The thermostat turns on the heating system or air conditioning when the room temperature strays outside the level you set, and turns it off when the temperature is correct. An electronic thermostat *(page 23)* can be programmed to turn the system on and off at certain times of day, or to change the preset temperature levels automatically. Small batteries run its display functions. The humidistat *(page 23)* senses moisture in the air to activate a humidifier or, in some cases, a dehumidifier, built into the heating system. It may be located on the wall, the air duct or the plenum.

Three typical thermostats are pictured below and at right. Those connected to central heating or cooling systems may be round or rectangular, and usually operate in a 24-volt circuit; the voltage rating is marked inside the cover or on the body. The household line voltage—120 or 240 volts—that powers the heating or cooling system is stepped down by a transformer before it passes to the thermostat circuit. Based on its setting and the room temperature, the thermostat signals various relays in its circuit to turn on or turn off heating elements,

gas valves, blowers and other components in the system. This chapter covers maintenance and repair of the thermostat wall unit; to troubleshoot other parts in the thermostat circuit or controlled by it, see the chapter on the particular system.

Most low-voltage thermostats use a temperature-sensitive bimetal coil that opens and closes a switch to turn the system on and off. The switch is commonly a mercury bulb type, in which a ball of mercury rolls toward or away from electrical contacts. Such a thermostat must be installed perfectly level. Heat pump systems with an auxiliary heating unit use a staged thermostat, which has two or three mercury bulb switches. It is serviced like a conventional thermostat.

The line-voltage thermostat *(right, bottom)* is wired to an electric baseboard heater that runs on 240-volt household line voltage. (Some heaters run on 120 volts.) Since voltage to the thermostat is not stepped down by a transformer, it is especially important to turn off power to the heater at the service panel, and test that it is off *(page 22)*, before touching the wires. To troubleshoot the heater itself, or a built-in thermostat, consult Baseboard Heaters *(page 84)*.

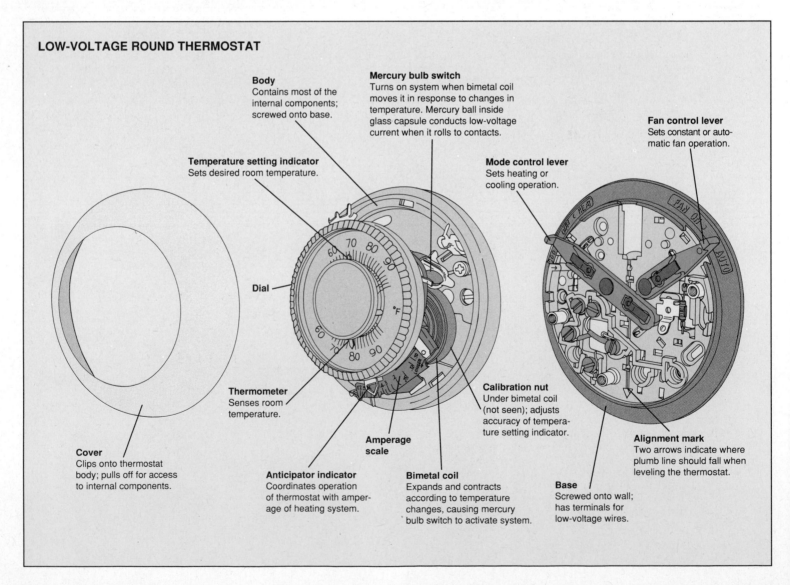

LOW-VOLTAGE ROUND THERMOSTAT

Body
Contains most of the internal components; screwed onto base.

Mercury bulb switch
Turns on system when bimetal coil moves it in response to changes in temperature. Mercury ball inside glass capsule conducts low-voltage current when it rolls to contacts.

Fan control lever
Sets constant or automatic fan operation.

Temperature setting indicator
Sets desired room temperature.

Mode control lever
Sets heating or cooling operation.

Dial

Calibration nut
Under bimetal coil (not seen); adjusts accuracy of temperature setting indicator.

Thermometer
Senses room temperature.

Alignment mark
Two arrows indicate where plumb line should fall when leveling the thermostat.

Amperage scale

Cover
Clips onto thermostat body; pulls off for access to internal components.

Anticipator indicator
Coordinates operation of thermostat with amperage of heating system.

Bimetal coil
Expands and contracts according to temperature changes, causing mercury bulb switch to activate system.

Base
Screwed onto wall; has terminals for low-voltage wires.

LOW-VOLTAGE RECTANGULAR THERMOSTAT

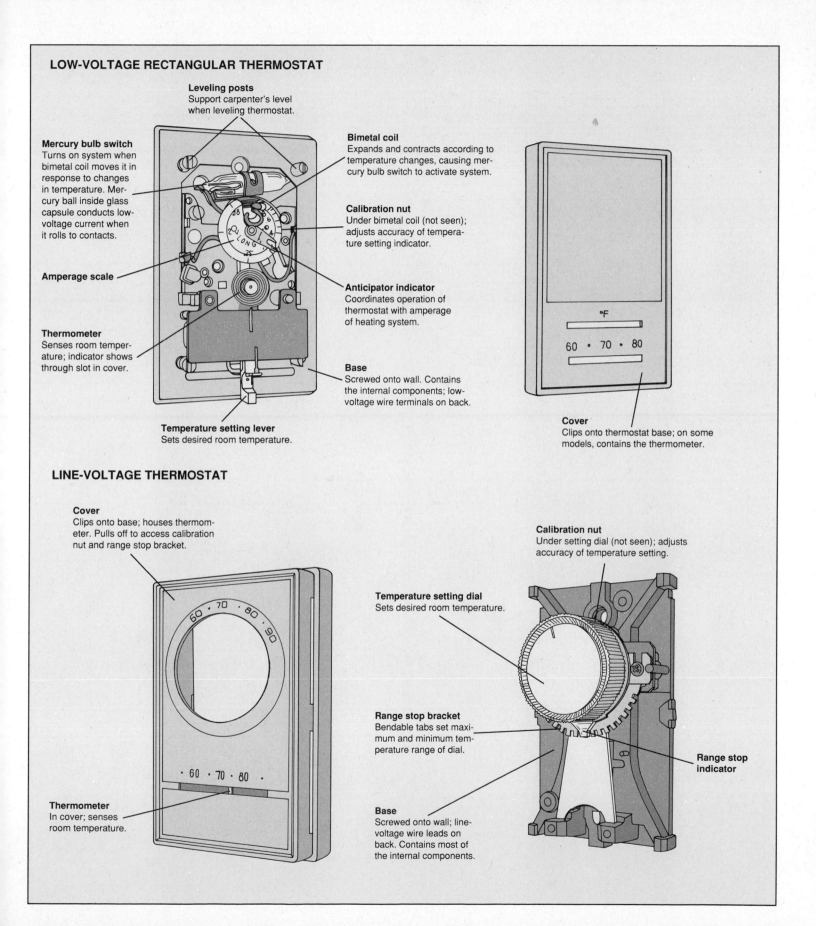

Leveling posts
Support carpenter's level when leveling thermostat.

Mercury bulb switch
Turns on system when bimetal coil moves it in response to changes in temperature. Mercury ball inside glass capsule conducts low-voltage current when it rolls to contacts.

Amperage scale

Thermometer
Senses room temperature; indicator shows through slot in cover.

Bimetal coil
Expands and contracts according to temperature changes, causing mercury bulb switch to activate system.

Calibration nut
Under bimetal coil (not seen); adjusts accuracy of temperature setting indicator.

Anticipator indicator
Coordinates operation of thermostat with amperage of heating system.

Base
Screwed onto wall. Contains the internal components; low-voltage wire terminals on back.

Temperature setting lever
Sets desired room temperature.

Cover
Clips onto thermostat base; on some models, contains the thermometer.

LINE-VOLTAGE THERMOSTAT

Cover
Clips onto base; houses thermometer. Pulls off to access calibration nut and range stop bracket.

Thermometer
In cover; senses room temperature.

Calibration nut
Under setting dial (not seen); adjusts accuracy of temperature setting.

Temperature setting dial
Sets desired room temperature.

Range stop bracket
Bendable tabs set maximum and minimum temperature range of dial.

Range stop indicator

Base
Screwed onto wall; line-voltage wire leads on back. Contains most of the internal components.

TROUBLESHOOTING GUIDE

SYMPTOM	POSSIBLE CAUSE	PROCEDURE
No heat	No power to system	Replace fuse or reset circuit breaker *(p. 134)* □○
	Thermostat dirty	Clean thermostat *(p. 18)* □○
	Thermostat faulty	Test thermostat *(low-voltage and line-voltage, p. 22* ◼️◓▲ ; *electronic, p. 23* □○*)*
	Electronic thermostat batteries weak or dead	Replace batteries *(p. 23)* □○
Heating exceeds or does not reach desired temperature	Thermostat not level	Level thermostat *(p. 21)* □○
	Anticipator set incorrectly	Adjust anticipator *(p. 19)* □○
	Thermostat out of calibration	Recalibrate low-voltage or line-voltage thermostat *(p. 20)* ◼️◓▲
	Range stop on line-voltage thermostat set too low or too high	Adjust range stops *(p. 21)* □○▲
Heating system short cycles (turns on and off repeatedly)	Thermostat dirty	Clean thermostat *(p. 18)* □○
	Anticipator set incorrectly	Adjust anticipator *(p. 19)* □○
Heating system does not turn off	Thermostat not level	Level thermostat *(p. 21)* □○
Cooling system does not turn on	No power to system	Replace fuse or reset circuit breaker *(p. 134)* □○
	Thermostat dirty	Clean thermostat *(p. 18)* □○
	Thermostat faulty	Test thermostat *(low-voltage and line-voltage, p. 22* ◼️◓▲ ; *electronic, p. 23* □○*)*
	Electronic thermostat batteries weak or dead	Replace batteries *(p. 23)* □○
Cooling exceeds or does not reach set temperature	Thermostat out of calibration	Recalibrate low-voltage or line-voltage thermostat *(p. 20)* ◼️◓▲
	Thermostat not level	Level thermostat *(p. 21)* □○
Humidifier does not turn on	No power to system	Replace fuse or reset circuit breaker *(p. 134)* □○
	Humidistat faulty	Test humidistat *(p. 23)* □○

DEGREE OF DIFFICULTY: □ **Easy** ◼️ **Moderate** ◼️ **Complex**
ESTIMATED TIME: ○ **Less than 1 hour** ◓ **1 to 3 hours** ● **Over 3 hours**   ▲ **Special tool required**

ACCESSING AND CLEANING THE THERMOSTAT COMPONENTS

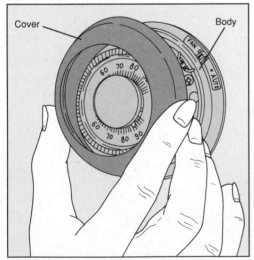

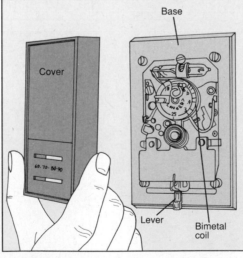

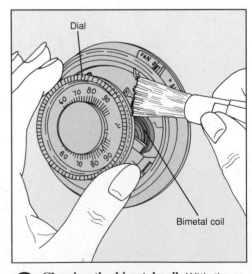

1 **Removing the cover.** Turn off power to the heating system at the main service panel *(page 134)*. Access the bimetal coil, anticipator, mercury bulb switch and lever contacts by snapping the cover of a round thermostat off its body *(above, left)* or a rectangular thermostat off its base *(above, right)*.

2 **Cleaning the bimetal coil.** With the thermostat cover removed, use a clean, soft brush to wipe built-up dust and dirt off the bimetal coil. Turn a round thermostat's dial *(above)* or a rectangular thermostat's lever from the lowest to the highest setting to help dislodge stubborn particles. If your thermostat is round, go to step 3; if it is rectangular, go to step 4.

ACCESSING AND CLEANING THE THERMOSTAT COMPONENTS (continued)

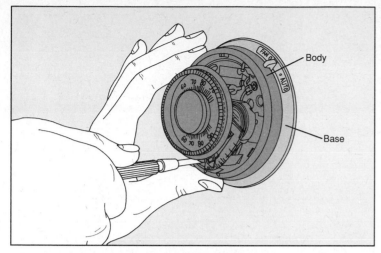

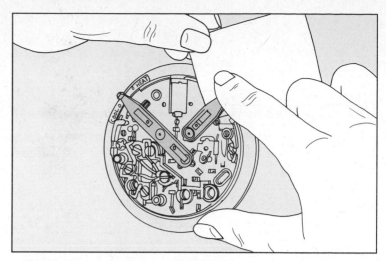

3 **Removing the body.** Use a screwdriver to loosen the screws securing a round thermostat's body to its base *(above)*; on many models, the screws will remain attached to the body. Take care not to remove recessed screws that secure electrical components. Remove the body and set it aside.

4 **Cleaning the switch contacts.** With a round thermostat's cover and body removed, or with a rectangular thermostat's cover removed, lift the lever and slip a strip of white bond paper between it and the contacts *(above)*. Shift the lever across the contacts and slide the paper to clean them. Tighten any loose wire connections. Remount a round thermostat body, replace the thermostat cover and turn on the heating system.

ADJUSTING THE ANTICIPATOR

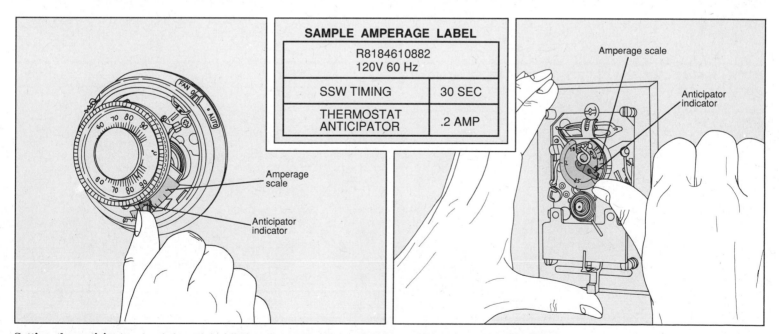

SAMPLE AMPERAGE LABEL	
R8184610882 120V 60 Hz	
SSW TIMING	30 SEC
THERMOSTAT ANTICIPATOR	.2 AMP

Amperage scale

Anticipator indicator

Amperage scale

Anticipator indicator

Setting the anticipator. Look for a label listing the amperage setting for your thermostat circuit *(inset)*: in your system's instruction manual, on an electric furnace's service panel, or on an oil- or gas-fired furnace or boiler's transformer or relay box. Turn off power to the heating system at the main service panel *(page 134)*, and remove the thermostat cover *(page 18)*. Check the position of the anticipator indicator on its scale. On an air distribution system, the indicator should point to the recommended amperage setting; on a water distribution system, the indicator should point to a number 1.4 times the amperage setting. If the anticipator is not correctly set, use your fingernail or a pen point to adjust the indicator; along a round thermostat's linear scale *(above, left)* or a rectangular thermostat's circular scale *(above, right)*. A two-stage thermostat has two adjustable anticipators; adjust each according to the recommended amperage setting for its heating unit.

RECALIBRATING A LOW-VOLTAGE THERMOSTAT

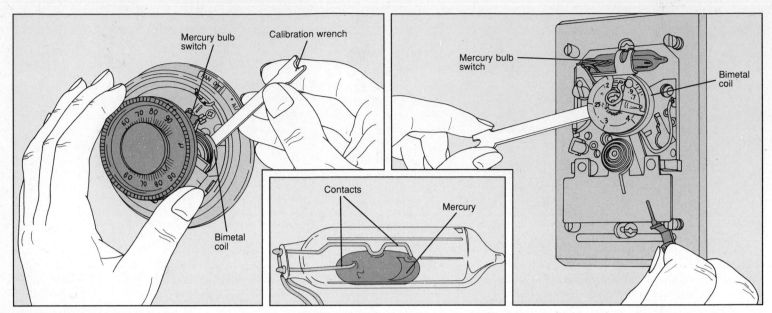

Recalibrating the thermostat. Turn off power to the heating system at the main service panel *(page 134)*. Pull off the thermostat cover *(page 18)* and check the calibration: Set the dial or lever at the actual room temperature; use a room thermometer to check the accuracy of the thermostat's thermometer. If this is the exact setting at which the mercury *(inset)* shifts from one end of the bulb to the other, the thermostat is properly calibrated. To recalibrate, set the thermostat 5°F above room temperature. Look for the calibration nut under the bimetal coil. Fit a calibration wrench *(page 132)* onto the nut and hold the thermostat dial or lever firmly. Do not touch, or breathe on, the heat-sensitive bimetal coils. Slowly rotate the nut, clockwise on most thermostats *(above, left and right)*, just until the mercury rolls to the right of the contacts. Turn the temperature setting down 10°F and wait five minutes, then set the thermostat dial or lever to match the room temperature reading. Holding the thermostat dial or lever firmly, turn the calibration nut in the opposite direction just until the mercury rolls back to the contacts. Allow the recalibrated thermostat to stabilize for 30 minutes. Check whether the room temperature matches the set temperature, and recalibrate again, if necessary. Reinstall the cover and turn on the power.

RECALIBRATING A LINE-VOLTAGE THERMOSTAT

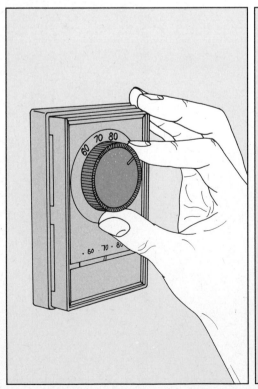

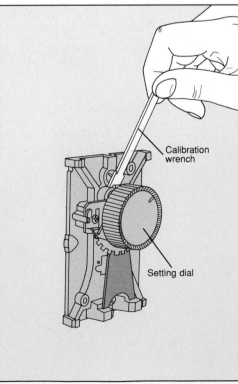

Recalibrating the thermostat. Turn up the setting dial *(far left)* just until you hear the thermostat click on. If the setting matches the temperature reading on the thermometer in the cover plate or on a room thermometer, the thermostat is calibrated properly. Otherwise, write down the two numbers, then turn off power to the heating system at the main service panel *(page 134)*. Rotate the setting dial to its highest setting and pull off the thermostat cover *(page 18)*. Fit a calibration wrench *(page 132)* onto the calibration screw behind the setting dial *(near left)*. Refer to the numbers you recorded: If the setting was higher than the temperature reading, rotate the calibration nut one-eighth turn clockwise for each degree F of difference; if the temperature reading was higher, rotate the wrench one-eighth turn counterclockwise for each degree F of difference. Replace the thermostat cover and rotate the setting dial to normal room temperature. Allow the thermostat to stabilize for 10 minutes, then check the calibration again; it may require several adjustments. Once finished, apply a tiny drop of nail polish to secure the calibration screw.

SETTING RANGE STOPS ON A LINE-VOLTAGE THERMOSTAT

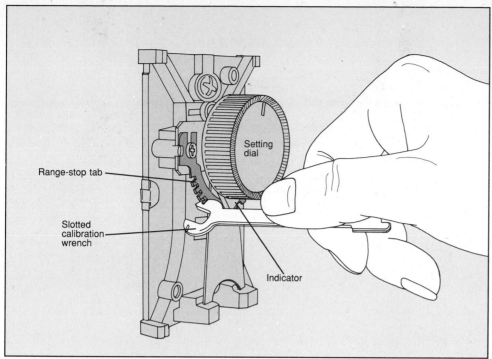

Adjusting the range stops. If your thermostat has range stops, check the settings: Turn the setting dial clockwise as far as it will go; the high-range stop is properly adjusted if the indicator stops at the highest desired setting. Then turn the setting dial counterclockwise as far as it will go; the low-range stop is properly adjusted if the indicator stops at the lowest desired setting. To adjust a range stop, turn off power to the heating system at the main service panel *(page 134)* and remove the thermostat cover *(page 18)*. Use a slotted calibration wrench or long-nose pliers to bend down the misadjusted high- or low-range stop tab. To set a new high-range stop, first reinstall the cover and move the setting dial to the highest desired temperature setting. Then remove the cover and bend up the range stop tab just to the left of the indicator. To adjust the low-range stop, move the setting dial to the lowest desired temperature setting, then remove the cover and bend up the range stop tab just to the right of the indicator. Reinstall the cover and turn on power to the heating system.

LEVELING A LOW-VOLTAGE THERMOSTAT

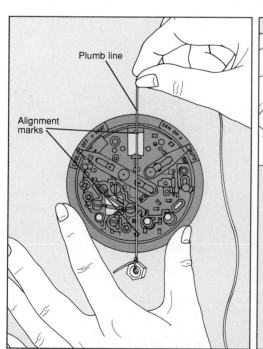

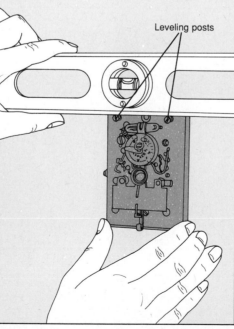

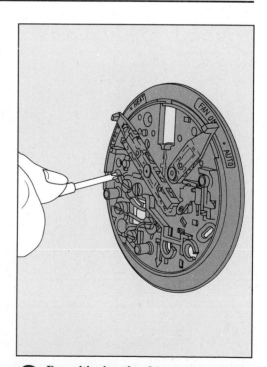

1 **Checking the level.** Turn off power to the heating system at the main service panel *(page 134)*, then remove a round thermostat's cover and body, or a rectangular thermostat's cover *(page 18)*. To check the level of a round thermostat, first make a plumb line by tying a nut to a short length of string. Suspend the string in front of the thermostat base *(above, left)*. The upper and lower alignment marks should both line up with the string; if not, loosen the screws *(step 2)* and reposition the thermostat. Check a rectangular model by resting a carpenter's level across the leveling posts at the top of the thermostat base *(above, right)*; the bubble should move to the center of the level. If not, reposition the thermostat *(step 2)*.

2 **Repositioning the thermostat.** Use a screwdriver to loosen the screws securing the circular *(above)* or rectangular thermostat's base to the wall. Level the base using the plumb line or the carpenter's level, then tighten the screws. Reassemble the thermostat and turn on power to the heating system.

TESTING AND REPLACING A LOW-VOLTAGE THERMOSTAT

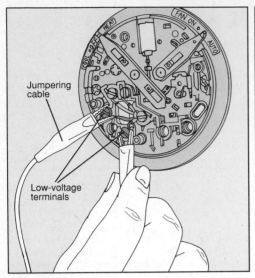

Jumpering cable

Low-voltage terminals

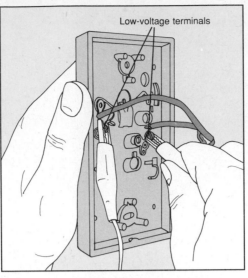

Low-voltage terminals

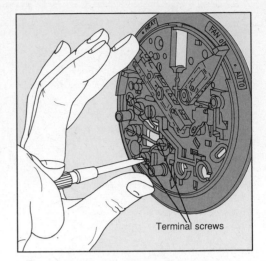

Terminal screws

1 **Jumpering the low-voltage terminals.** Turn off power to the heating system at the main service panel *(page 134)*. Set the thermostat to AUTO and HEAT. Remove a round thermostat's cover and body *(page 18)*; unscrew a rectangular thermostat's base from the wall. Using a jumpering cable *(page 132)*, attach the alligator clips to the terminals marked "R" and "W" connected to the red wire and white wire from the wall: on the front of a round thermostat *(above, left)* or the back of a rectangular thermostat *(above, right)*. Turn on the power. If the furnace or boiler still does not turn on, the thermostat tests OK; turn off the power and reassemble the thermostat. If the furnace or boiler now turns on, the thermostat is faulty; turn off the power.

2 **Replacing the thermostat.** Disconnect the wires from their terminals on the base; do not let them slip into the wall. Feed the wires through the opening in a new round thermostat's base, and connect them *(above)*; on a rectangular model, connect the wires to the terminals on its back. Screw the new thermostat in place, then install the body on a round style. Level the thermostat *(page 21)*. Snap on the cover and turn on the power.

TESTING AND REPLACING A LINE-VOLTAGE THERMOSTAT

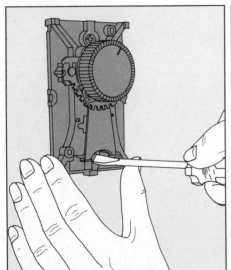

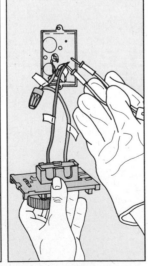

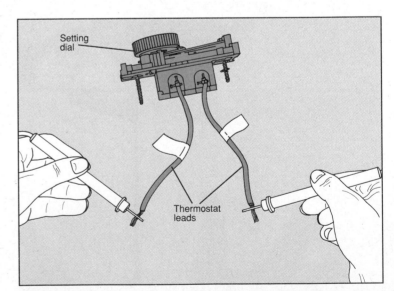

Setting dial

Thermostat leads

1 **Freeing the thermostat.** Turn off power to the heating system at the main service panel *(page 134)*. Unscrew the thermostat *(above, left)* and gently pull it out of the wall box to expose the wires. Label all wires *(page 135)*. Taking care not to touch any bare wires, unscrew one wire cap. Use a voltage tester to check that power is off: Wearing an insulated glove, hold the tester in one hand; touch one probe to the uncapped wires and the other probe to the metal wall box *(above, right)*. Twist off the other wire cap; repeat the test. Finally, touch a probe to each uncapped wire connection. In each case, the tester should not light. If it does, flip off the proper circuit breaker; repeat the voltage test.

2 **Testing at the wire leads.** Turn the setting dial from the lowest to the highest setting, and listen for a click at the room temperature setting. If it does not click, set a multitester to RX1K to test continuity *(page 136)*. Turn the setting dial to the lowest setting, and touch a multitester probe to each thermostat lead *(above)*; the multitester should not show continuity. Next, turn the dial past room temperature and test again; the tester should show continuity. If the thermostat tests OK, reconnect the leads, reassemble the thermostat and turn on the power. If it fails either test or does not click, replace it with a new thermostat of equal or greater amperage rating.

SERVICING AND TESTING AN ELECTRONIC THERMOSTAT

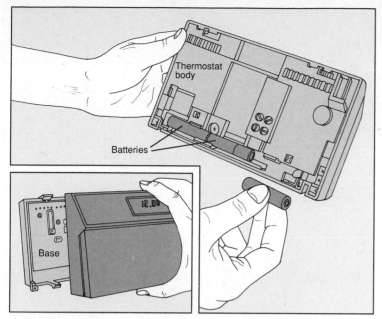

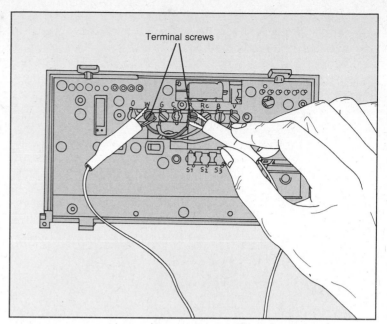

1 **Access to the internal components.** Turn off power to the heating system at the main service panel *(page 134)*. Grasp the thermostat body and pull the unit off its base *(inset)* to expose the battery compartment and terminals. If the low-battery signal or indicator light is on, remove the batteries and replace them with new identically-rated alkaline batteries *(above)*. Re-install the thermostat and restore power to the heating system. If the batteries are OK, go to step 2.

2 **Jumpering the low-voltage connections.** With the power off and the thermostat body removed *(step 1)*, look at the base and locate the two terminal screws (marked "R" and "W") connected to the red wire and white wire from the wall. Using a jumpering cable *(page 132)*, attach one alligator clip to each terminal *(above)*. Turn on the power. If the furnace or boiler now starts, the thermostat is faulty and should be replaced. Buy an identical thermostat; mount it on the original base if desired.

TESTING AND REPLACING A HUMIDISTAT

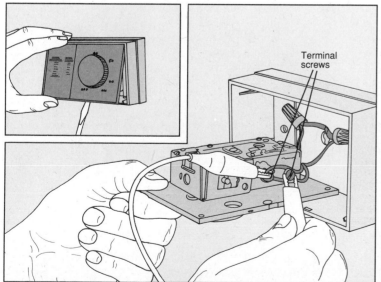

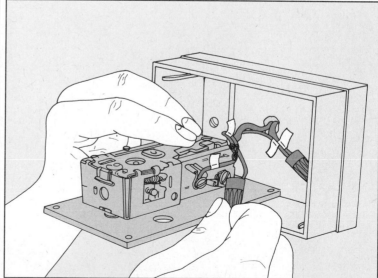

Accessing and testing the humidistat. For valid test results, the room humidity must be lower than the humidistat setting. Turn off power to the heating system at the main service panel *(page 134)*. Insert a screwdriver in the access slot to pry open the humidistat body *(inset)*, and gently pull the humidistat out of its housing. Locate the two wires running to the internal humidistat assembly. Using a jumpering cable *(page 132)*, attach one alligator clip to each of the two terminal screws *(above, left)*. Turn on power to the heating system; if the humidifier drum now rotates, the humidistat is faulty; turn off the power. Label the wires and remove the wire caps *(above, right)* to disconnect the humidistat; replace it with an identical unit, available from a heating supplies dealer.

AIR DISTRIBUTION

An air distribution system can provide the home with an ideal year-round climate. It circulates heated or cooled air through a gas, oil or electric furnace, or central air conditioning unit, via a network of ducts in the home. A blower, housed in the furnace and turned by a belt-drive or direct-drive motor, draws air from the return duct, through the filter, and into the furnace, where the air is heated or cooled. The air then goes out through the supply duct to branch ducts. On many systems, adjustable dampers in the ducts regulate the flow of air to different parts of the house; adjustable registers regulate air flow into a room. Gas and oil furnaces generally use a fan-

and-limit control mounted on the plenum to switch the blower on and off in response to plenum temperature. Electric furnaces *(page 76)* and central air conditioning *(page 102)* use separate blower relays or fan centers to switch the blower in direct response to the thermostat.

A humidifier mounted directly on the supply duct responds to a humidistat, and runs only when the furnace operates. Most models pull air through a wet pad, evaporating moisture to increase the amount of water vapor in the air. The evaporator tray is filled with water automatically, through a supply tube. An adjustable float regulates the water level.

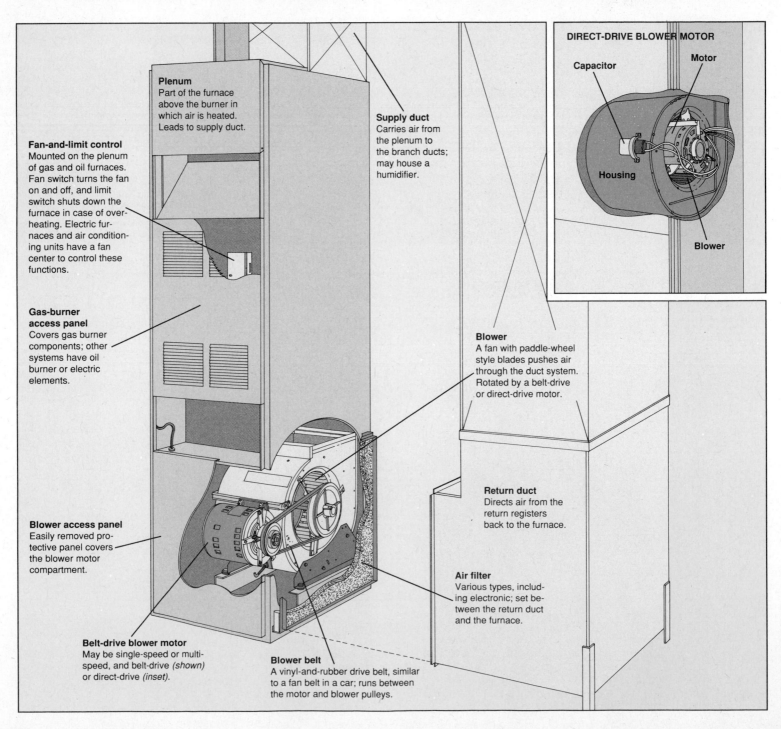

DIRECT-DRIVE BLOWER MOTOR

Capacitor

Motor

Housing

Blower

Plenum
Part of the furnace above the burner in which air is heated. Leads to supply duct.

Supply duct
Carries air from the plenum to the branch ducts; may house a humidifier.

Fan-and-limit control
Mounted on the plenum of gas and oil furnaces. Fan switch turns the fan on and off, and limit switch shuts down the furnace in case of overheating. Electric furnaces and air conditioning units have a fan center to control these functions.

Gas-burner access panel
Covers gas burner components; other systems have oil burner or electric elements.

Blower
A fan with paddle-wheel style blades pushes air through the duct system. Rotated by a belt-drive or direct-drive motor.

Return duct
Directs air from the return registers back to the furnace.

Blower access panel
Easily removed protective panel covers the blower motor compartment.

Air filter
Various types, including electronic; set between the return duct and the furnace.

Belt-drive blower motor
May be single-speed or multispeed, and belt-drive *(shown)* or direct-drive *(inset)*.

Blower belt
A vinyl-and-rubber drive belt, similar to a fan belt in a car; runs between the motor and blower pulleys.

If properly maintained, an air distribution system will be relatively trouble-free. At least once a year, clean the blower blades *(page 30)* and lubricate the blower motor *(page 27)*; inspect a belt-drive motor's belt, and check its tension *(page 28)*. The duct system should last the life of the house without cleaning or repair. Once a month during the heating and cooling seasons, check the air filter, and clean or replace it as necessary *(page 26)*. A humidifier filter and evaporator tray should be cleaned monthly as well *(page 36)*.

Troubleshooting an air distribution system may be as simple as cleaning the blower blades *(page 30)* or a more complicated procedure, such as replacing a direct-drive blower motor *(page 35)*. Follow all safety instructions in this chapter and in the Emergency Guide *(page 8)*. Before beginning most repairs, turn off power to the system at the main service panel *(page 134)* and unit disconnect switch *(page 135)*.

Heating and cooling problems may be caused by another component in the system. Also consult the Troubleshooting Guides in System Controls *(page 16)*, Oil Burners *(page 52)*, Gas Burners *(page 68)*, Electric Furnaces *(page 76)*, Heat Pumps *(page 90)* and Central Air Conditioning *(page 102)*, where appropriate.

TROUBLESHOOTING GUIDE

SYMPTOM	POSSIBLE CAUSE	PROCEDURE
Heating or central air conditioning does not run at all	No power to unit	Reset circuit breaker or replace fuse *(p. 134)* □○
System runs, but no hot or cold air flow	Blower belt broken or loose	Inspect and adjust belt tension *(p. 28)* □○ and pulley alignment *(p. 29)* □○; replace belt *(p. 28)* ◨○
	Fan-and-limit control faulty	Test fan-and-limit control *(p. 31)* ◨○▲; replace *(p. 32)* ◨◑ if necessary
	Blower relay faulty	Test blower relay *(p. 32)* ◨◑▲; replace if necessary
	Blower motor faulty	Inspect blower motor *(p. 31)* □◑; discharge, test and replace capacitor *(p. 34)* ◨◑▲; replace belt-drive blower motor *(p. 33)* ◨◑ or direct-drive blower motor *(p. 35)* ■● , if necessary
Heating or cooling insufficient in part of the house	Register closed or blocked, or filter dirty	Open register and remove obstructions; replace dirty register filter
	Blower blades dirty	Clean blower blades *(p. 30)* □○
	Duct disconnected or damper closed	Reconnect duct or open damper
Heating or cooling insufficient throughout entire house	Air filter dirty	Clean or replace air filter *(p. 26)* □○
	Blower running poorly	Oil blower motor and bearings *(p. 27)* □○; adjust blower speed *(p. 27)* □○; adjust blower belt tension *(p. 28)* □○ and blower pulley alignment *(p. 29)* □○; clean blower blades *(p. 30)* □○
	Blower bearings dry or worn	Lubricate bearings *(p. 27)* □○; replace bearings *(p. 29)* □○
	Duct dampers adjusted improperly	Adjust dampers
Air flow noisy	Blower speed too high	Adjust blower speed *(p. 27)* □○
	Joints loose at duct corners	Tighten joints and adjust duct hangers, or call for service to rebuild duct joints
Blower noise excessive	Blower motor or blower bearings dry or worn	Lubricate motor and bearings *(p. 27)* □○; replace blower bearings *(p. 29)* ◨◑; replace belt-drive motor *(p. 33)* ◨◑ or direct-drive motor *(p. 35)* ■●
	Motor or blower mounting hardware loose	Tighten mounting hardware *(pp. 29, 33, 35)* □○
Blower belt slips or squeals	Belt tension incorrect	Adjust belt tension or replace belt *(p. 28)* □○
	Pulley misaligned	Align blower pulleys *(p. 29)* ◨◑
HUMIDIFIER		
Air too dry	Evaporator pad dirty or clogged	Wash or replace pad *(p. 36)* □○
	Humidistat faulty	Test humidistat *(System Controls, p. 23)* ◨◑
	Humidifier motor faulty	Test motor *(p. 38)* ◨◑▲; replace if necessary
	Transformer faulty	Test transformer *(p. 38)* ◨◑▲; replace if necessary
	Water level too low	Adjust water level float *(p. 37)* □○
Water overflowing or leaking	Water level too high	Adjust water level float *(p. 37)* □○

DEGREE OF DIFFICULTY: □ **Easy** ◨ **Moderate** ■ **Complex**

ESTIMATED TIME: ○ **Less than 1 hour** ◑ **1 to 3 hours** ● **Over 3 hours**

▲ **Special tool required**

SERVICING FILTERS

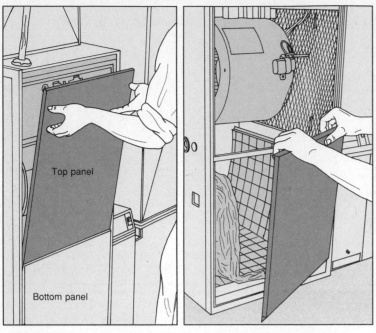
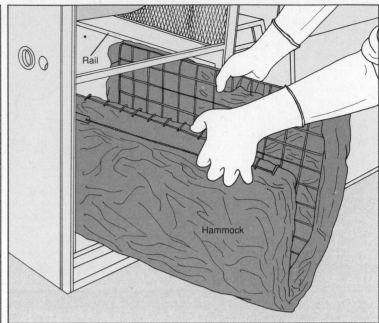

Replacing a hammock filter. Turn off power to the furnace at the main service panel *(page 134)* and unit disconnect switch *(page 135)*. Access the filter by removing the furnace cover panels *(above, left and center)*, releasing each with a sharp upward tug. Once a month during the heating and cooling seasons, inspect the filter by shining a light through it from behind. If the light is nearly or completely blocked, replace the filter. Lift and release each side of the wire mesh hammock from the retaining rails, fold it inward, and pull the filter and hammock out of the furnace *(above, right)*. Hammock filters made of fiberglass batting are available in rolls from a heating and cooling supplies dealer. Use a utility knife to cut the batting to size, place it between the holders and slide the hammock onto the retaining rails. Reinstall the access panels and restore power to the furnace.

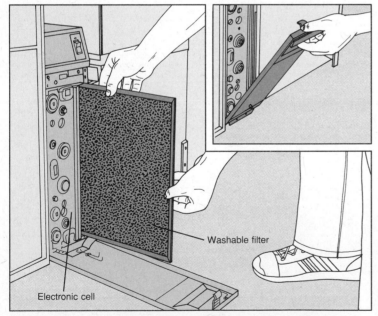
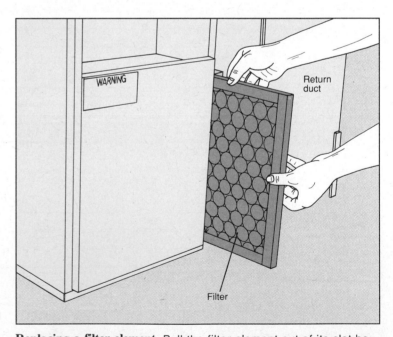

Cleaning an electronic filter. Turn off power to the furnace at the main service panel *(page 134)* and unit disconnect switch *(page 135)*. Pull open the filter door panel *(inset)*. Slide the filter out of its retainer sleeve *(above)*; push aside any spring tabs that secure the filter in place. Once a month during heating and cooling seasons, wash the filter with a detergent-and-water solution, rinse well, then allow it to drip dry before reinserting. If the filter is damaged, replace it with an identical one. Before each season, slide out the electronic cells and carefully wash them the same way. Close the door panel and turn on the power.

Replacing a filter element. Pull the filter element out of its slot between the return duct and the blower *(above)*. On some units, the filter is inside the furnace; turn off power to the furnace at the main service panel *(page 134)* or unit disconnect switch *(page 135)* and pull off any access panels. Remove the filter and inspect it for dirt by holding it up to a light bulb. If the light is nearly or completely blocked, replace a dirty cardboard-framed fiberglass filter with an exact replacement; wash a dirty metal or plastic element filter using a hose with a high-pressure nozzle, then let it drip dry. Reinstall the filter and turn on the power.

LUBRICATING THE BLOWER AND MOTOR

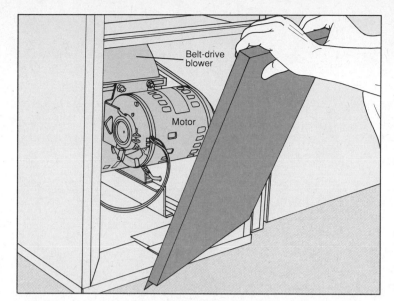

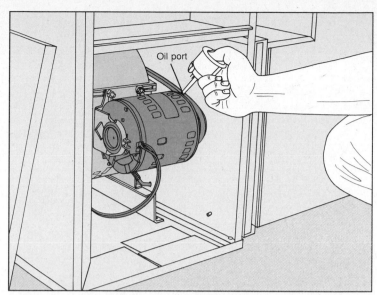

1 **Accessing the blower.** Turn off power to the furnace at the main service panel *(page 134)* and unit disconnect switch *(page 135)*. The blower, located in the bottom of the unit, is covered by an access panel that usually has a label with oiling instructions. Tug up and out to remove the panel *(above)*. Most blowers with oil ports should be lubricated at least once a year. Often only the motor requires oiling; most belt-drive blowers turn on sealed bearings that should not be oiled at all. Direct-drive blowers turn on the motor's bearings.

2 **Lubricating the bearings.** Locate the motor-bearing oil ports, if any, at each end of the motor casing. If the ports are capped, use the blade of a flat-tipped screwdriver to pry up the caps. Squirt five drops of non-detergent SAE-30 motor oil into each port *(above)*; do not over-oil. If there are oil ports at the bearings where the blower shaft is mounted, oil them also.

ADUSTING THE BLOWER SPEED

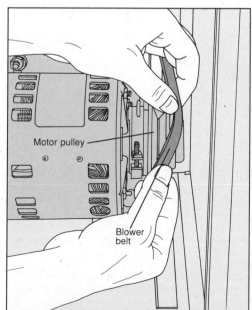

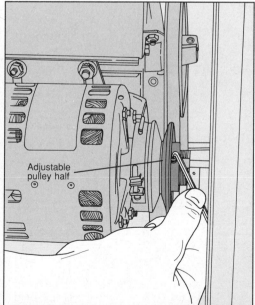

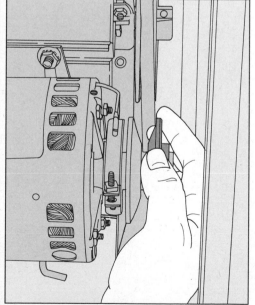

Adjusting the pulley width. The speed at which the adjustable motor pulley drives the blower belt is determined by the width set between the two halves of the pulley. Turn off power to the furnace at the main service panel *(page 134)* and unit disconnect switch *(page 135)*; access the blower *(step above)*. Slip the blower belt over the outer lip of the motor pulley *(above, left)*; turn the pulley counterclockwise to help work it off. Use a hex wrench to loosen the pulley setscrew *(above, center)* until the adjustable pulley half can be turned on its threaded shaft. Turn the pulley one-half turn at a time: clockwise to increase blower speed and counterclockwise to decrease it *(above, right)*. Align the setscrew with the flat side of the motor shaft before retightening it. Slip the belt back on the pulley, and check its tension *(page 28)*.

ADJUSTING THE BLOWER BELT TENSION

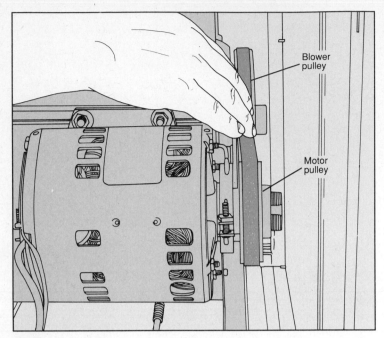

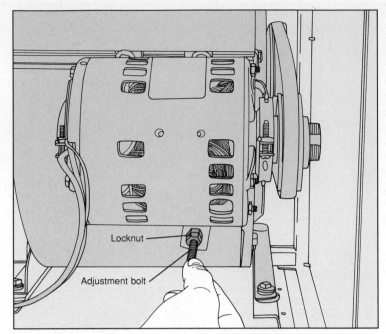

1 **Checking the belt tension.** Turn off power to the furnace at the main service panel *(page 134)* and unit disconnect switch *(page 135)*; remove the access panels *(page 27)*. Inspect the belt for cracks, brittleness or wear; replace it if damaged *(below)*. Test belt tension by pushing down on the belt midway between the pulleys until the belt is taut *(above)*; it should slacken about one inch. If it slackens more, it should be tightened or replaced, depending on your model; look for instructions on a label near the blower. If it slackens less than one inch, it is too tight; loosen it.

2 **Adjusting the belt tension.** With power to the furnace turned off, locate the locknut on the blower-belt adjustment bolt and loosen it with an open-end wrench. Next, turn the adjustment bolt by hand, clockwise to tighten the belt, or counterclockwise to loosen it *(above)*. Check the belt tension *(step 1)*, then retighten the locknut. Reinstall the access panel and turn on power to the furnace.

REPLACING THE BLOWER BELT

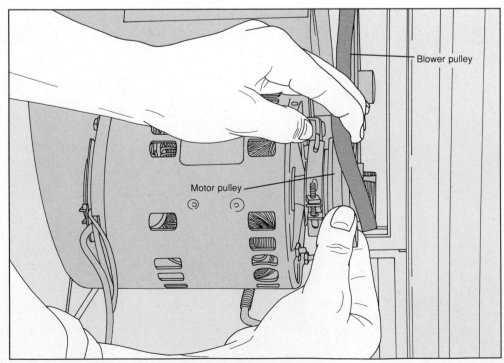

Installing a new blower belt. Turn off power to the furnace at the main service panel *(page 134)* or unit disconnect switch *(page 135)* and remove the access panels covering the blower *(page 27)*. Inspect the blower belt; if it is cracked, brittle, worn or stretched beyond adjustment limits, replace it. Remove the belt by pushing it over the top of the outer lip of the motor pulley with your thumb; if the belt is hard to remove, use your other hand to turn the pulley counterclockwise. Take the old belt with you to a heating supplies dealer and buy an exact replacement. Place the new belt over the blower pulley. Hold the belt in the top groove of the motor pulley with one hand and turn the pulley counterclockwise with the other hand *(left)*. Reassemble the unit and restore the power.

ALIGNING THE BLOWER AND MOTOR PULLEYS

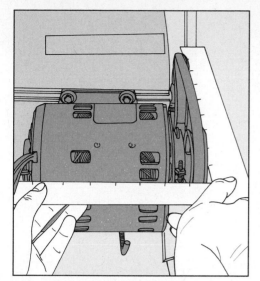

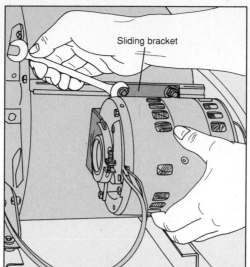

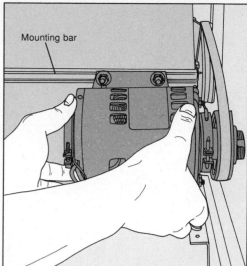

1 **Checking pulley alignment.** Turn off power to the furnace at the service panel *(page 134)* and unit disconnect switch *(page 135)* and remove the access panels *(page 27)*. Hold a carpenter's square or straightedge across the faces of the motor and blower pulleys *(above)*. If the edge does not rest flush against both pulleys, shift the motor's position on its bracket until the pulleys align *(step 2)*.

2 **Shifting the motor.** Use an open-end wrench to loosen the nuts securing the sliding bracket to the mounting bar *(above, left)*. Grasp the motor in both hands and shift it slightly forward or backward, as needed *(above, right)*. Recheck the pulley alignment with the carpenter's square *(step 1)*; when the pulleys are aligned properly, retighten the mounting bolts. Restore power to the furnace and check that the belt runs smoothly and quietly.

REPLACING THE BLOWER BEARINGS

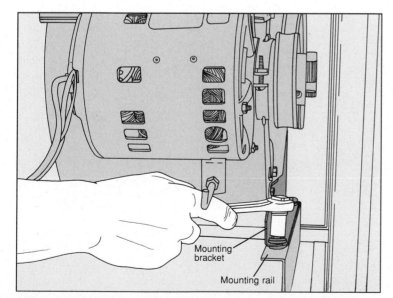

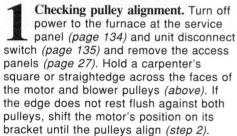

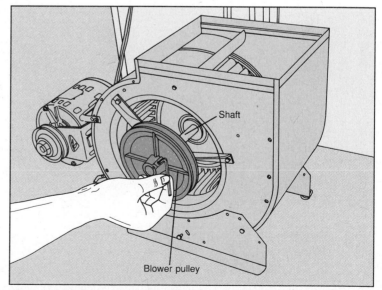

1 **Unbolting the blower assembly.** Turn off power to the furnace at the main service panel *(page 134)* and unit disconnect switch *(page 135)*. Remove the access panels and take off the blower belt *(page 27)*. Turn the blower pulley by hand to check the bearings; grating noises, roughness or wobbling indicate worn bearings. Using a socket wrench, unscrew all the bolts holding the blower's mounting brackets to the mounting rails *(above)*. Slide the blower assembly out of the furnace.

2 **Removing the blower pulley.** Insert a hex wrench into the blower pulley setscrew and turn it counterclockwise to loosen the blower pulley *(above)*. If the setscrew is difficult to turn, apply penetrating oil and try again after 10 minutes. Pull the blower pulley off the shaft.

REPLACING THE BLOWER BEARINGS (continued)

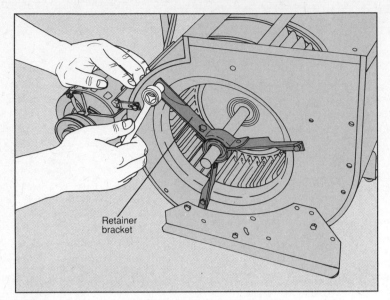

Retainer
bracket

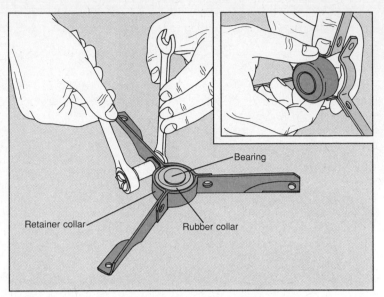

Bearing

Retainer collar

Rubber collar

3 **Removing the bearing retainer bracket.** The blower has bearings on both sides, one at each end of the shaft. Use a socket wrench or nut driver to remove the bolts holding the bearing retainer brackets to the blower housing *(above)*. Slide the brackets off both ends of the shaft. Take the old bearings to a heating supplies dealer and purchase exact replacements.

4 **Replacing a bearing.** Using two wrenches, remove the nut and bolt from the retainer collar that grips the bearing *(above)*. Bend the collar apart with your fingers or with a flat-tipped screwdriver and take out the bearing *(inset)*. Slide off the rubber collar around the bearing and keep it for installation around the new bearing. Install the bearing in the bracket and reassemble the blower. Reassemble the unit and restore the power.

CLEANING THE BLOWER BLADES

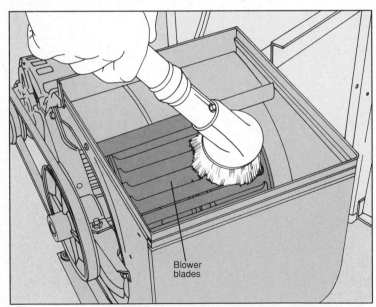

Blower
blades

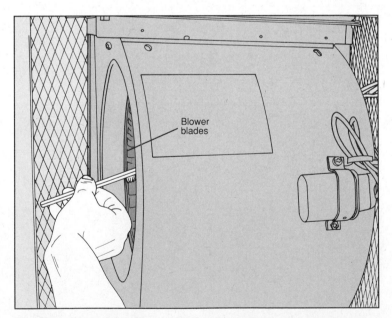

Blower
blades

Cleaning belt-drive blower blades. Blower blades load up with dirt and become inefficient only if the filter has been neglected in exceptionally dusty conditions. Turn off power to the furnace at the main service panel *(page 134)* and unit disconnect switch *(page 135)*. Remove the blower assembly from the furnace *(page 29)* to access the blower blades. Vacuum the blades with a brush attachment *(above)*. If necessary, remove the belt *(page 28)* and blower pulley *(page 29)* to insert a small brush inside to loosen dirt. Tap the blower casing to dislodge the dirt. Replace the blower assembly, reassemble the unit and restore the power.

Cleaning direct-drive blower blades. Turn off power to the furnace at the main service panel *(page 134)* and unit disconnect switch *(page 135)* and remove the access panels covering the blower *(page 27)*. Reach into the blower cavity with a toothbrush to clean the inside and outside surfaces of each blade. If a vacuum nozzle will fit, use the vacuum to dislodge dirt. If the blades are inaccessible, remove the direct-drive blower for cleaning *(page 35)*. After cleaning the blades, reassemble the unit and restore the power.

TESTING THE FAN-AND-LIMIT CONTROL

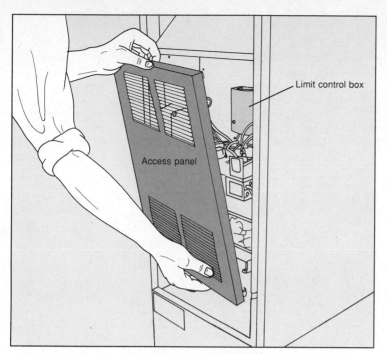

1 Accessing the control box. The fan-and-limit control used on gas and oil furnaces consists of a fan (blower) switch and a limit (shutoff) switch. On gas furnaces, the box is mounted near other controls: behind the access panel and against the plenum. Lift off the panel *(above)* and set it aside. An oil furnace fan-and-limit control is usually mounted directly on the plenum.

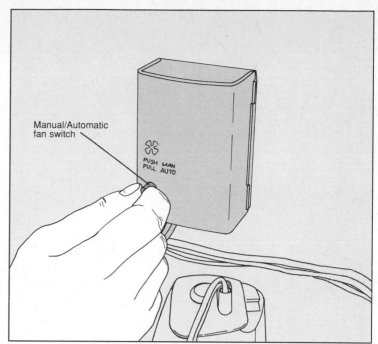

2 Testing the blower motor. Look on the limit control box for a push/pull or toggle switch marked MANUAL/AUTOMATIC or SUMMER/WINTER, or a switch mounted separately on an exterior panel. Set the button or toggle to MANUAL or SUMMER *(above)*. If the blower runs, the blower motor is OK. If the blower does not run, test the fan switch *(step 3)*.

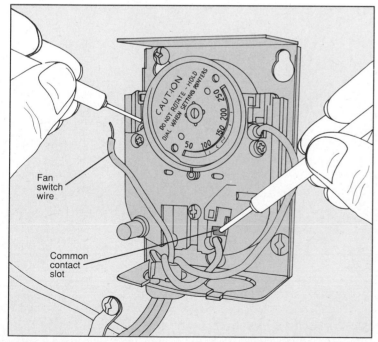

3 Testing the fan switch. Turn off power to the furnace at the main service panel *(page 134)* and unit disconnect switch *(page 135)*. Pull off the control cover. Disconnect the fan wire from its terminal. Set a multitester to measure continuity *(page 136)*. With the switch on MANUAL, touch one probe to the fan contact slot and the other probe to the common contact slot *(above)*. If there is continuity, the fan switch is OK; go to step 2. If there is no continuity, replace the fan-and-limit control *(page 32)*.

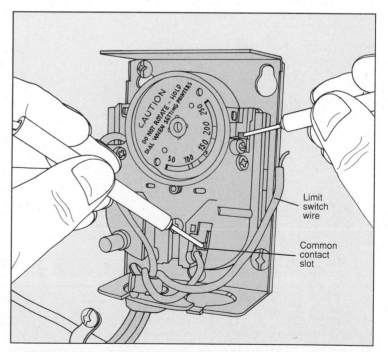

4 Testing the limit switch. Set a multitester to the RX1K setting to test continuity *(page 136)*. Disconnect the limit switch wire from its terminal. Touch one tester probe to the limit switch contact slot and the other probe to the common contact slot *(above)*. If there is no continuity, the limit control is faulty; replace it. Otherwise, reassemble the fan-and-limit control.

REPLACING THE FAN-AND-LIMIT CONTROL BOX

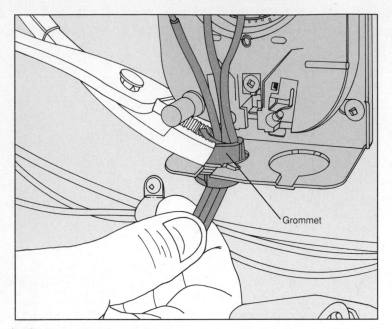

Grommet

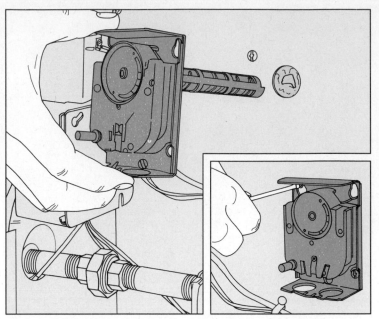

1 **Removing the wiring.** Turn off power to the furnace at the main service panel *(page 134)* and unit disconnect switch *(page 135)*. Pinch the inside rim of the wire grommet with pliers and push it out through the wiring access hole in the bottom of the fan-and-limit control box *(above)*.

2 **Unfastening a fan-and-limit control.** Label the wires *(page 135)* and disconnect them. Use a screwdriver to remove the screws holding the fan-and-limit control on the plenum *(inset)*. Pull the control out of the furnace *(above)*. Install an identical replacement, reversing these steps. Reassemble the furnace and restore power to the system.

TESTING AND REPLACING THE BLOWER RELAY

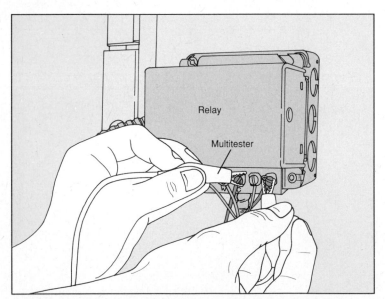

Relay

Multitester

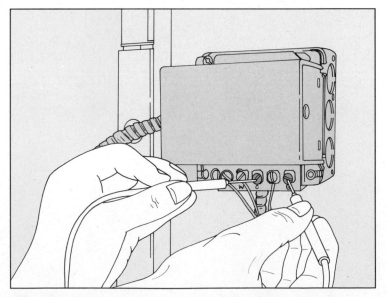

1 **Testing the thermostat circuit.** Turn off power to the heating and cooling system at the main service panel *(page 134)* and unit disconnect switch *(page 135)*. Set a multitester to the ACV setting, 50-volt range, to test voltage *(page 137)*. Attach one tester clip to the G terminal, and the other clip to the C, T or V terminal *(above)*. Without touching the clips or terminals, have someone turn on the power and turn the thermostat fan switch to ON. The tester should read about 24 volts and the blower should click on. If not, the thermostat circuit is faulty; see System Controls *(page 16)*. If the reading is 24 volts, turn off the power and go to step 2.

2 **Testing the relay for continuity.** With power to the system off, set a multitester to test continuity *(page 136)*. Touch one probe to the G terminal and the other to the C terminal *(above)*. There should be continuity. If there is no continuity, replace the relay *(step 3)*. If there is continuity through the relay, but the blower will not run, check the blower motor *(page 33)*.

TESTING AND REPLACING THE BLOWER RELAY (continued)

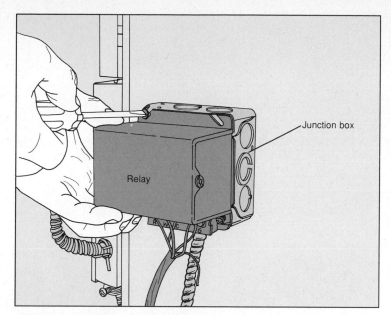

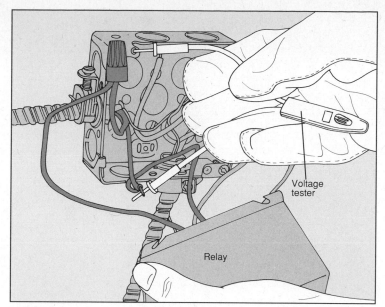

3 **Accessing the blower leads.** The wire leads from the blower enter the junction box and are connected to wires at the back of the relay. With power to the furnace off, remove the retaining screws with a screwdriver *(above)*, and pull the relay off the junction box. **Caution:** Check that power is off before disconnecting any wires *(step 4)*.

4 **Replacing the relay box.** Taking care not to touch any bare wires, uncap the black and white wires in the relay box. Holding a voltage tester *(page 132)* in one hand, touch one probe to the black wire ends and the other to the grounded junction box *(above)*. Next, test between the white wires and the box. Finally, test between the black and white wires. The tester should not glow in any case; if it does, turn off the correct circuit and test again. Label all wires for correct reconnection *(page 135)*, and replace the relay with an identical new one. Reconnect the wires, reassemble the unit and turn on the power.

TESTING AND REPLACING A BELT-DRIVE BLOWER MOTOR

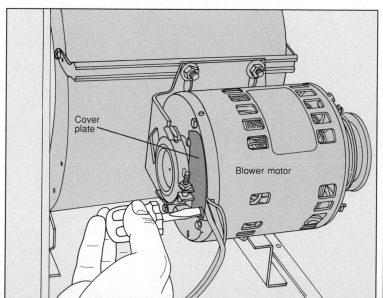

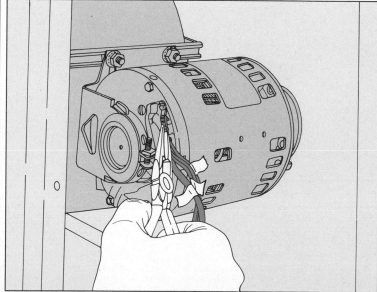

1 **Checking the motor.** Turn off power to the furnace at the main service panel *(page 134)* and unit disconnect switch *(page 135)*. Access the motor *(page 27)*. Try turning the motor shaft; if it moves stiffly, lubricate the motor *(page 27)*. If it will not turn at all, the motor is seized; replace it. Slip off the blower belt *(page 27)*, then use a short screwdriver to remove the cover plate screws *(above, left)*. Pull off the cover plate to expose the motor wiring. Label the wires for correct reconnection *(page 135)*, grip the spade lug connectors with long-nose pliers and pull them off the terminals *(above, right)*.

TESTING AND REPLACING A BELT-DRIVE BLOWER MOTOR (continued)

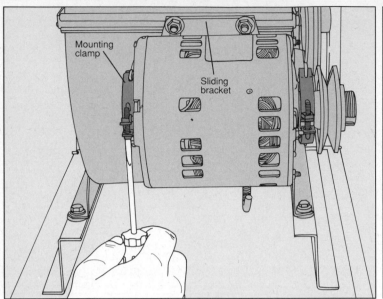

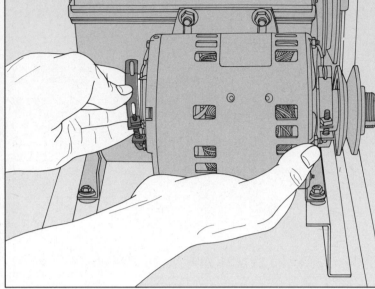

2 **Removing the motor.** Loosen or remove the mounting clamp screws *(above, left)*. Supporting the motor in one hand, unhook and remove the clamps *(above, right)* to free the motor. Unscrew the pulley *(page 27)* and keep it for reinstallation on the new motor. Install a new motor by reversing these instructions. Do not remove the sliding bracket from the blower housing unless the pulley installed on the new motor requires alignment *(page 29)*.

DISCHARGING AND TESTING A BLOWER MOTOR CAPACITOR

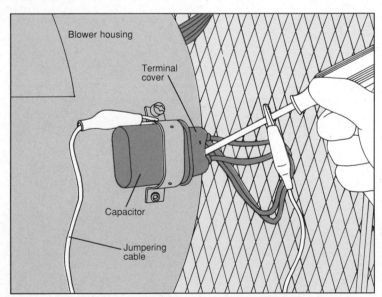

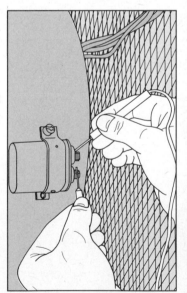

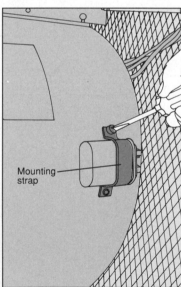

1 **Discharging the capacitor.** Turn off power to the furnace at the service panel *(page 134)* and unit disconnect switch *(page 135)*. Remove the blower access panels *(page 26)*. **Caution:** Do not attempt any repair until the capacitor has been discharged. Make a capacitor discharging tool *(page 140)*. To discharge the capacitor, clip one end of the capacitor discharger to an unpainted metal part on the unit chassis. Hold the insulated screwdriver handle of the discharger in one hand, reach through the capacitor cover with the tip of the screwdriver, and touch each capacitor terminal for one second *(above)*.

2 **Testing and replacing the capacitor.** With power off and the capacitor discharged, set a multitester to the RX1K scale to measure resistance *(page 136)*. Slip the terminal cover off the capacitor. Label the wires *(page 135)* and use long-nose pliers to pull the connectors off the terminals. Touch one multitester probe to each capacitor terminal *(above, left)*. The multitester needle should swing to the right, then sweep back. If the multitester needle does not move, or moves and does not sweep back, replace the capacitor: Loosen the screws on its mounting strap *(above, right)* and slide out the capacitor. Replace it with an identically-rated capacitor, reconnect the wires and restore the power.

REPLACING A DIRECT-DRIVE BLOWER MOTOR

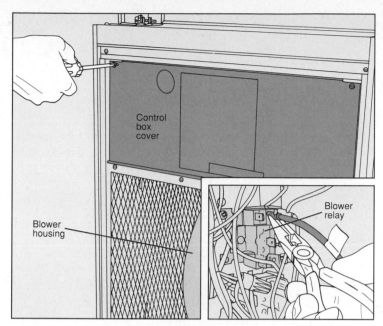

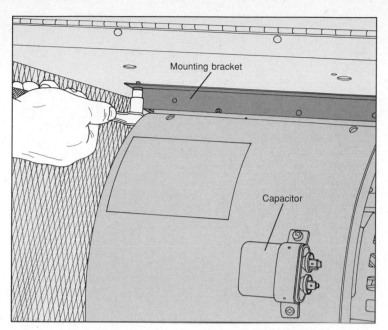

1 Accessing the blower wire terminals. Turn off power to the furnace at the main service panel *(page 134)* and unit disconnect switch *(page 135)*. Remove access panels to the blower *(page 27)*. Locate the point at which the motor's wire leads can be disconnected: at the motor, the relay and switch terminals or—on an electric furnace—at the control box *(above)*. Label the wires *(page 135)*, then use long-nose pliers to disconnect them *(inset)*.

2 Unfastening the blower. Discharge and disconnect the capacitor *(page 34)*. Use a socket wrench to remove the retaining bolts from the mounting bracket *(above)*. (The weight of the blower is held by the slide rails; it will not fall.) If the motor is mounted in the bottom of the furnace, unbolt it from its slide rails *(page 29)*.

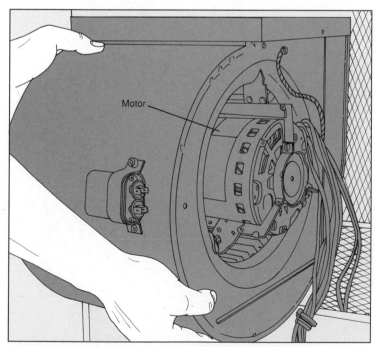

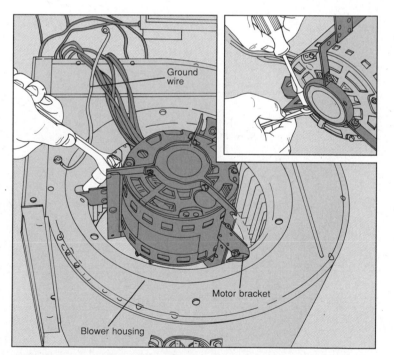

3 Checking and removing the blower. Try turning the motor shaft; if it moves stiffly, lubricate the motor *(page 27)*. If it will not turn at all, the motor is seized; replace it. A direct-drive blower and motor weighs 25 to 50 pounds; have a helper remove it with you. Crouching down, support the blower housing firmly and slide it out until it clears the slide rails *(above)*. Place the assembly motor-side up on the floor.

4 Replacing the motor. Disconnect the ground wire from the motor. Turn the blower on its side and, using a hex wrench, loosen the setscrew to release the motor shaft from the blower fan. Turn the blower upright and, using a socket wrench, unbolt the motor bracket from the blower housing *(above)*. Pull the motor out of the blower and place it on the floor; unscrew the bracket collar bolts using a screwdriver and open-end wrench *(inset)*. Purchase an identical replacement from a motor shop or from the manufacturer, and install it by reversing these instructions.

MAINTAINING A HUMIDIFIER

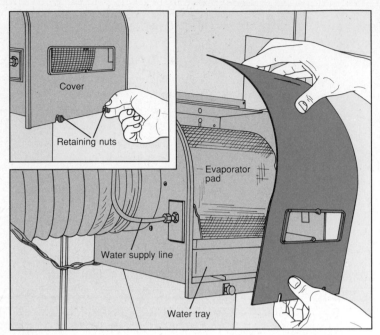

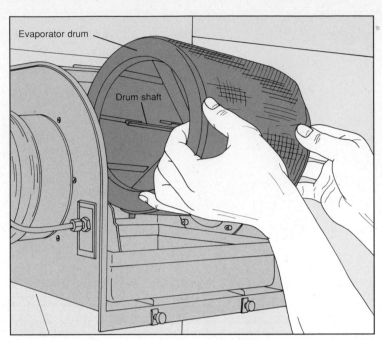

1 **Accessing the evaporator pad and tray.** At the end of each heating season, replace the evaporator pad, turn off the valve on the water line that supplies the humidifier, and clean the water tray. Turn off power to the heating and cooling system at the main service panel *(page 134)* and unit disconnect switch *(page 135)*. Loosen the retaining nuts along the bottom lip of the humidifier cover *(inset)* and lift off the cover to access the evaporator pad and water tray *(above)*.

2 **Removing the evaporator drum.** Grasp both ends of the evaporator drum and lift the shaft out of its slots *(above)*. Water mineral content varies by region; inspect the pad monthly during the heating season until you can identify how often your pad requires cleaning. When the pad is no longer soft and spongy, remove it *(step 3)* and clean or replace it.

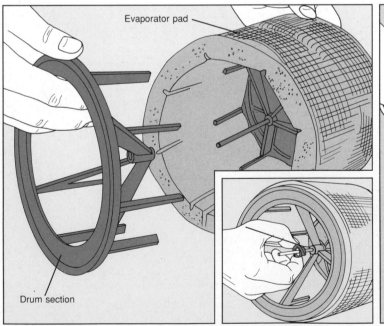

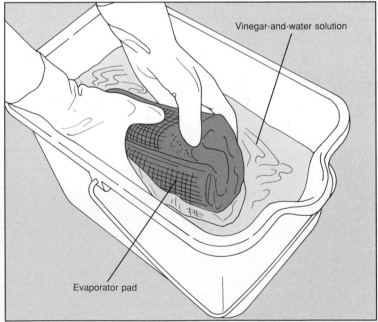

3 **Cleaning or replacing the evaporator pad.** Pinch the drum-shaft retaining clip and pull it off the drum shaft *(inset)*. Pull apart the two sections of the drum to release the pad *(above, left)*. If the pad is slightly hardened, soak it in a solution of three parts vinegar to one part water until it softens, then squeeze the solution through it to wash it out *(above, right)*. If the pad does not soften or has deteriorated, replace it with an identical new one, available from a hardware store or a heating and cooling supplies dealer.

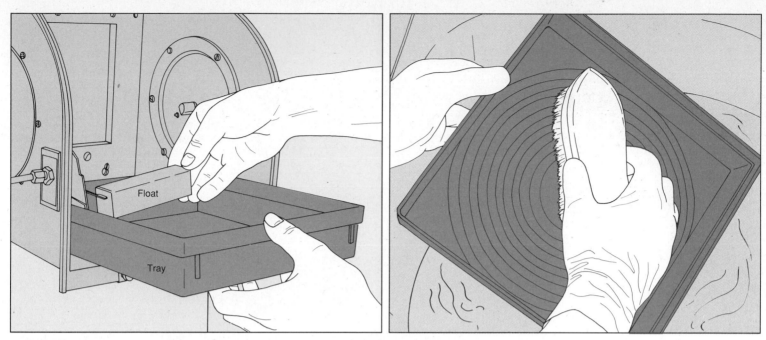

4 **Cleaning the tray.** With the water line valve turned off, lift the float and pull out the tray *(above, left)*. To dislodge scale deposits, wash the tray using a stiff brush and a vinegar-and-water solution or humidifier descaler—available from a hardware store *(above, right)*. Rinse the tray well. Add water-treatment tablets or liquid—also available from a hardware store—to inhibit mineral buildup and bacteria; follow instructions on the package. Reassemble the humidifier and turn on the water valve; restore the power.

ADJUSTING THE HUMIDIFIER WATER LEVEL

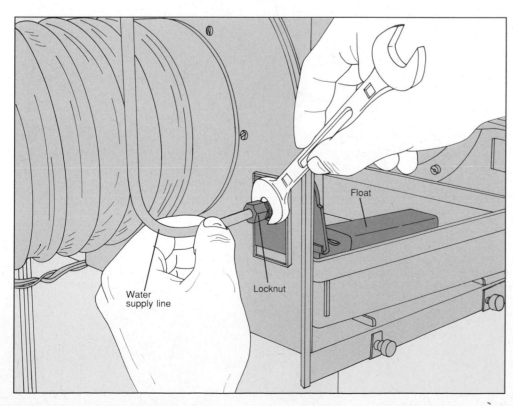

Adjusting the water level. Turn off power to the system at the main service panel *(page 134)* and unit disconnect switch *(page 135)*. Remove the cover panel *(page 36, step 1)* and check the depth of water in the tray. The water should be about 1 1/2 inches deep, enough to soak the evaporator pad as it turns. To adjust the water level, loosen the float assembly locknut on the water supply line with an open-end wrench *(left)*. Then move the float up to raise the water level, or down to lower the water level. Retighten the locknut securely against the retainer plate. Check the new water level. Reinstall the humidifier cover and restore the power.

TESTING AND REPLACING A HUMIDIFIER MOTOR

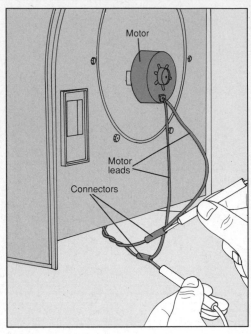

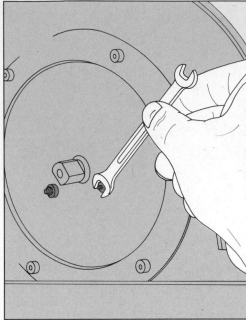

Testing and replacing the motor. Turn off power to the system at the main service panel *(page 134)* and unit disconnect switch *(page 135)*. Set a multitester *(page 136)* to the ACV setting, 50-volt range. To test whether power is reaching the motor, insert a multitester probe into each connector between the low-voltage leads and motor leads *(far left)* or, if necessary, remove the connectors to test the low-voltage leads. Being careful not to touch the tester or any bare wires, turn on power to the furnace, making sure the blower is running, and turn the humidistat to its highest setting. If the multitester shows no voltage, test the transformer *(step below)*. If the multitester registers voltage, power is reaching the motor; if the motor doesn't run, it is faulty. To replace it, turn off power and remove the cover and drum *(page 36)*. From inside, use a wrench to remove the motor's mounting nuts *(near left)*. Pull the motor out of its bracket and install an identical replacement. Connect the new motor leads to the low-voltage leads. Reassemble the humidifier and restore power.

TESTING AND REPLACING A HUMIDIFIER TRANSFORMER

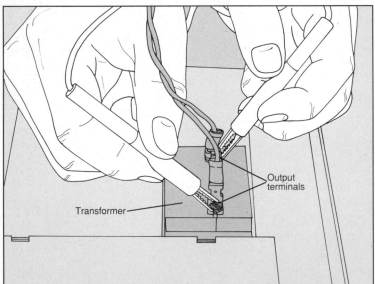

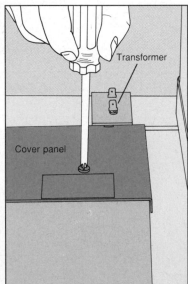

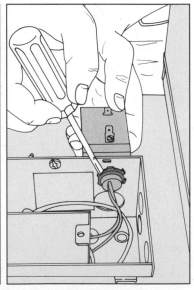

Testing and replacing the transformer. Turn off power to the system at the main service panel *(page 134)* and unit disconnect switch *(page 135)*. Trace the low-voltage wires from the humidifier motor to the transformer. Set a multitester *(page 136)* to the ACV setting, 50-volt range. Clip a probe to each transformer terminal *(above, left)*. Being careful not to touch the tester or any bare wires, turn on power to the system, making sure the blower is running, and set the humidistat to its highest setting. The multitester should register voltage. If the multitester registers voltage at the transformer terminals but not at the motor connections *(step above)*, the low-voltage wiring may be faulty; call for service. If there is no voltage, the transformer is faulty; replace it. Turn off the power. **Caution:** Check to confirm that power is indeed off before disconnecting any wires, as for a relay *(page 33, step 4)*. Being careful not to touch any bare wire ends, uncap the wires in the transformer box. Holding a voltage tester in one hand *(page 132)*, touch one probe to the black wire ends and the other to the grounded metal box. Next, test between the white wires and the metal box. Finally, test between the black and white wire ends. The tester should not glow in any case, confirming that the power is off. Disconnect the wires. Use a screwdriver to push the transformer locknut counterclockwise to remove it *(above, right)*. Lift out the transformer and replace it with an identically-rated unit. Reconnect the wires, reassemble the unit and restore the power.

WATER DISTRIBUTION

Most water distribution systems found in homes today were installed before the 1950s. In fact, many boilers still in use are converted coal-burning units, like the one pictured below.

When the thermostat calls for heat, the burner turns on to heat the water inside the boiler. When the water reaches a preset temperature (usually 100°F to 120°F), the circulator pump turns on, and circulates the heated water throughout the house via a system of pipes to radiators or convectors. One or more aquastats monitor the water temperature. The burner aquastat acts as a safety device, turning off the burner if the water exceeds a safe temperature. When the thermostat is satisfied, the burner stops. The pump aquastat turns the circulator pump on and off. Even with the burner off, the circulator pump continues to pump water through the system until the water cools to below the aquastat's preset temperature.

A pressure reducing valve reduces incoming water pressure of 70 to 75 psi (pounds per square inch) to the 15 psi minimum required by the boiler. Pressure within the boiler is registered on a pressure gauge. When the system is cold, the gauge should read about 15 psi, depending on the height of the radiators above the boiler. As the water heats up, pressure increases to about 20 psi; a safety valve on or near the boiler protects it from bursting by releasing water through a discharge pipe at 30 psi and above.

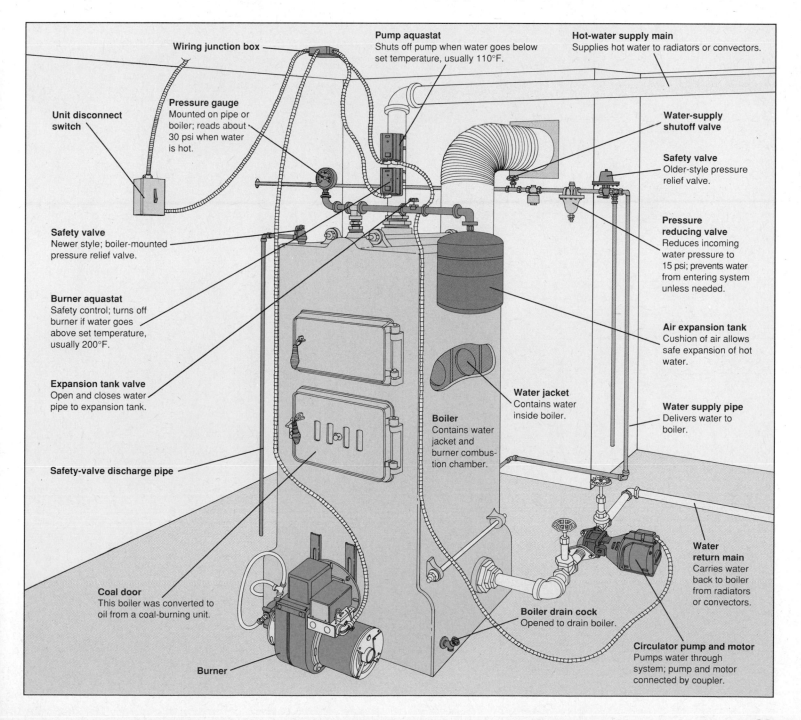

Wiring junction box

Pump aquastat
Shuts off pump when water goes below set temperature, usually 110°F.

Hot-water supply main
Supplies hot water to radiators or convectors.

Unit disconnect switch

Pressure gauge
Mounted on pipe or boiler; reads about 30 psi when water is hot.

Water-supply shutoff valve

Safety valve
Older-style pressure relief valve.

Safety valve
Newer style; boiler-mounted pressure relief valve.

Pressure reducing valve
Reduces incoming water pressure to 15 psi; prevents water from entering system unless needed.

Burner aquastat
Safety control; turns off burner if water goes above set temperature, usually 200°F.

Air expansion tank
Cushion of air allows safe expansion of hot water.

Expansion tank valve
Open and closes water pipe to expansion tank.

Water jacket
Contains water inside boiler.

Water supply pipe
Delivers water to boiler.

Boiler
Contains water jacket and burner combustion chamber.

Safety-valve discharge pipe

Coal door
This boiler was converted to oil from a coal-burning unit.

Boiler drain cock
Opened to drain boiler.

Water return main
Carries water back to boiler from radiators or convectors.

Burner

Circulator pump and motor
Pumps water through system; pump and motor connected by coupler.

Most systems use an air expansion tank, usually installed directly above or alongside the boiler; some large systems have two tanks. This tank contains a pocket of air that is compressed when the hot water expands. If an old-style tank becomes completely full of water it must be drained and recharged *(page 46)*; the newer diaphragm-type tanks become waterlogged only when the diaphragm is worn out, and the tank must be replaced *(page 46)*.

Before each heating season, bleed all the radiators or convectors in your home and replace any faulty bleed valves *(page 42)*. Check the pressure reducing valve *(page 43)*. Inspect the air expansion tank for leaks and recharge an old-style tank *(page 46)*. Test the aquastats *(page 47)* and oil the pump and motor *(page 49)*.

Before beginning most tests and repairs, turn off power to the boiler at the main service panel *(page 134)* and unit disconnect switch *(page 135)*. **Caution:** Working with wiring is highly dangerous if the proper electrical circuit is not turned off; test to make sure the power is off *(page 48)*. Identify the components of your system in advance. If necessary, ask a plumber to point them out.

Heating problems may be caused by another system unit. Consult the Troubleshooting Guides in System Controls *(page 16)*, Oil Burners *(page 52)* and Gas Burners *(page 68)*.

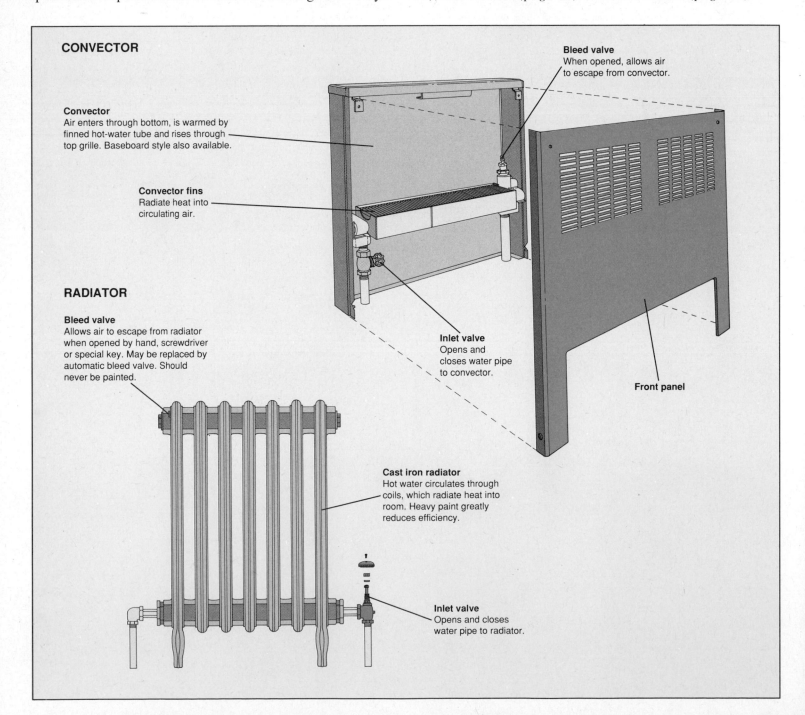

CONVECTOR

Bleed valve
When opened, allows air to escape from convector.

Convector
Air enters through bottom, is warmed by finned hot-water tube and rises through top grille. Baseboard style also available.

Convector fins
Radiate heat into circulating air.

RADIATOR

Bleed valve
Allows air to escape from radiator when opened by hand, screwdriver or special key. May be replaced by automatic bleed valve. Should never be painted.

Inlet valve
Opens and closes water pipe to convector.

Front panel

Cast iron radiator
Hot water circulates through coils, which radiate heat into room. Heavy paint greatly reduces efficiency.

Inlet valve
Opens and closes water pipe to radiator.

TROUBLESHOOTING GUIDE

SYMPTOM	POSSIBLE CAUSE	PROCEDURE
No heat	No power to system	Replace fuse or reset circuit breaker (p. 134) □○
	Aquastat faulty	Test aquastats (p. 47) □○ and replace (p. 48) ▣◓▲ if necessary
	Air trapped in radiators or convectors	Bleed radiators or convectors (p. 42) □○; if frequent bleeding is required, install automatic bleed valves (p. 43) ▣●
	Coupler broken	Replace coupler (p. 50) ■●▲
	Circulator pump faulty	Call for service
No heat, motor housing hot, burning odor	Circulator motor burned out	Test motor windings and replace motor (p. 50) ▣◓▲ if necessary
Gurgling sound in pipes or radiator	Air trapped in radiator	Bleed radiators (p. 42) □○; if frequent bleeding is required, install automatic bleed valves (p. 43) ▣◓
Too much heat, burner does not turn off	Burner aquastat faulty	Test aquastat (p. 47) □○ and replace (p. 48) ▣◓▲ if necessary
Heat uneven throughout house	Air trapped in radiator or convector	Bleed radiators or convectors (p. 42) □○; if frequent bleeding is required, install automatic bleed valves (p. 43) ▣◓
	Coupler broken	Replace coupler (p. 50) ■●▲
	Circulator pump faulty	Call for service
	Pressure reducing valve set too low	Call for service
Radiator is cold, hammering noise in pipes	Sagging floor makes radiator slope, trapping cold water or air pockets in radiator	Level sloping radiator with 1/4-inch wood shims, raising the leg below bleed valve
Not enough heat, convector is lukewarm	Convector fins dirty or bent	Vacuum convector fins, and straighten them with broad-billed pliers
	Air trapped in convector	Bleed convectors (p. 42) □○; if frequent bleeding is required, install automatic bleed valves (p. 43) ▣◓
	Circulator pump faulty	Call for service
Not enough heat, radiators are lukewarm (water pressure is too low)	Pressure gauge set too low	Call for service
	Pipes leaking	Tighten pipe joints with a wrench or call for service
	Pressure reducing valve faulty	Check valve (p. 43) ▣◓; if faulty, call for service
	Rust deposits in boiler or pipes	Drain and refill boiler system, adding rust inhibitor (p. 43) ▣●
	Circulator pump faulty	Call for service
Radiators are lukewarm (water pressure exceeds 30 psi soon after burner turns on, and safety valve discharges)	Old-style expansion tank waterlogged or leaking	Recharge expansion tank (p. 46) ▣●; if leaking, call for service
	Diaphragm-type expansion tank waterlogged or leaking	Replace expansion tank (p. 46) ▣◓
Radiators above boiler are lukewarm, others are cold	Pump aquastat faulty	Test aquastat (p. 47) □○ and replace (p. 48) ▣◓▲ if necessary
	Circulator motor burned out	Test motor windings and replace motor (p. 50) ▣◓▲ if necessary
	Circulator pump faulty or leaking	Call for service
Circulator motor noisy	Motor and pump not properly maintained	Lubricate circulator pump and motor (p. 49) □○
Circulator motor sounds like chain being dragged through system	Coupler broken	Replace coupler (p. 50) ■●▲
Water spills from safety-valve discharge pipe	Old-style expansion tank waterlogged or leaking	Recharge expansion tank (p. 46) ▣●; if leaking, call for service
	Diaphragm-type expansion tank waterlogged or leaking	Replace expansion tank (p. 46) ▣◓
Circulator pump leaks	Pump seal or impeller worn	Call for service
Inlet valve leaks	Valve packing deteriorated	Replace valve packing (p. 45) ▣◓
Bleed valve drips constantly or will not turn	Bleed valve faulty	Replace bleed valve (p. 42) ▣●

DEGREE OF DIFFICULTY: □ **Easy** ▣ **Moderate** ■ **Complex**
ESTIMATED TIME: ○ **Less than 1 hour** ◓ **1 to 3 hours** ● **Over 3 hours**

▲ **Special tool required**

BLEEDING THE SYSTEM

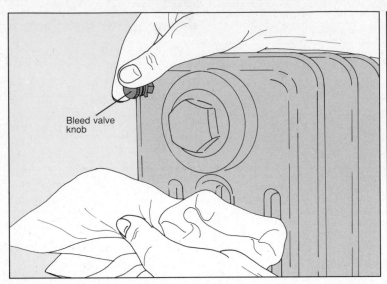
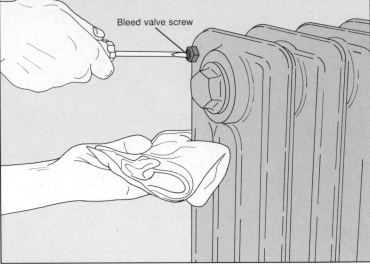

Bleed valve knob

Bleed valve screw

Bleeding a radiator or convector. At the beginning of each heating season, bleed the valve on each radiator or convector, starting on the top floor of your home. Have an absorbent rag handy; when the air is vented, water will escape from the valve. On models with a bleed-valve knob, turn the knob 180 degrees counterclockwise to purge the radiator or convector of air *(above, left)*. On models without a bleed-valve knob, use a screwdriver *(above, right)* or radiator venting key, as appropriate. Automatic bleed valves are supposed to bleed themselves, but water impurities can occasionally clog them. If a radiator with an automatic bleed valve is cool, bleed the valve: Take off the cap, turn it upside down and push it into the air valve, as for a car tire. On all bleed valves, as soon as water escapes in a steady stream the air has been bled; close the valve.

REMOVING AND REPLACING A BLEED VALVE

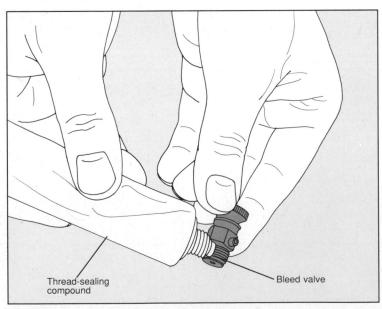
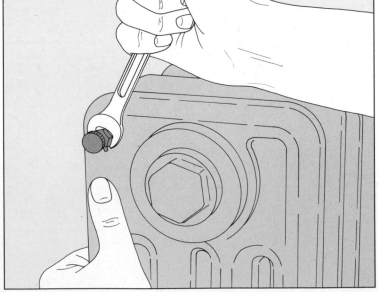

Thread-sealing compound

Bleed valve

Replacing a conventional valve. If a bleed valve drips constantly or will not bleed at all, replace it. Buy a new bleed valve to fit your radiator, and coat its threads with thread-sealing compound *(above, left)*. Turn off power to the boiler at the main service panel *(page 134)* and unit disconnect switch *(page 135)*; allow the boiler to cool. Drain the system *(page 44)* just until water stops leaking from the bleed valve. Use an open-end wrench to loosen the old bleed valve from its fitting *(above, right)*, and unscrew the valve. Screw the new valve in place, taking care not to cross-thread the fitting. Refill the system *(page 44)*, then restore power. If the new valve leaks, try tightening the fitting, or call for service. A conventional valve may be replaced by an automatic bleed valve *(next step)*.

REMOVING AND REPLACING A BLEED VALVE (continued)

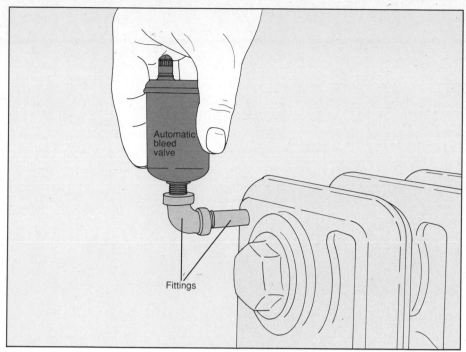

Installing an automatic bleed valve. If a radiator requires bleeding more often than twice a year, consider replacing its conventional bleed valve with an automatic valve. Buy an automatic bleed valve and the proper fittings for its upright, side-mounted or horizontal-mounted position. Coat all the threads with thread-sealing compound. Turn off power to the boiler at the main service panel *(page 134)* and unit disconnect switch *(page 135)*; allow the boiler to cool. Drain the system *(page 44)* just until water stops leaking from the bleed valve. Remove the old bleed valve *(page 42)*. Use a small pipe wrench to screw the automatic valve's fittings into the radiator, then screw on the new valve, taking care not to cross-thread the fittings *(left)*. Refill the system, carefully following all instructions *(page 44)*, and leave the bleed vent open. If the new bleed valve leaks, try tightening the fittings, or call for service.

CHECKING THE PRESSURE REDUCING VALVE

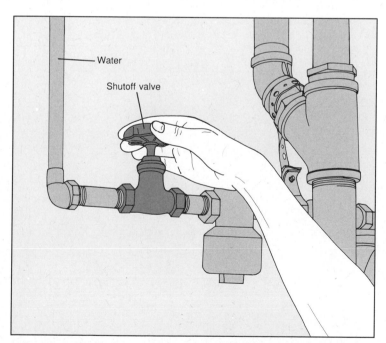

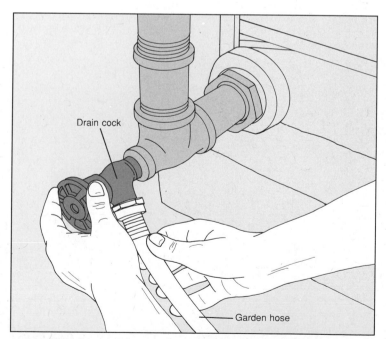

1 **Turning off water to the boiler.** Turn off power to the boiler at the main service panel *(page 134)* and unit disconnect switch *(page 135)*. Next, turn off the water supply at the shut-off valve on the water supply pipe leading to the boiler *(above)*. This pipe is usually the smallest one in the system, and leads to the pressure reducing valve. On some systems, it may be located behind the boiler. Wait for the boiler to cool, then drain the air expansion tank *(page 46)*.

2 **Testing the valve.** Locate the drain cock, near the bottom of the boiler, and screw on a garden hose long enough to reach the basement drain. (If no drain is available, direct the hose into a sink or out a window.) Let air into the system by opening one bleed valve *(page 42)* on the top floor. Then turn the drain cock handle counterclockwise to release about a bucketful of water. Close the drain cock and open the main shutoff valve; water should rush through the supply pipe to compensate for the drained water. If it does not, the pressure reducing valve may be faulty; have it checked by a professional.

DRAINING AND REFILLING THE SYSTEM

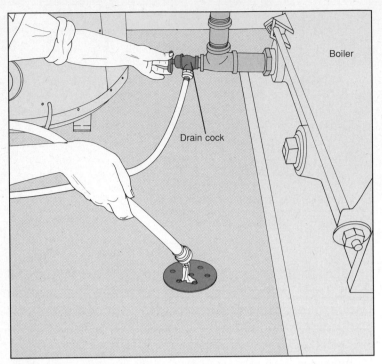

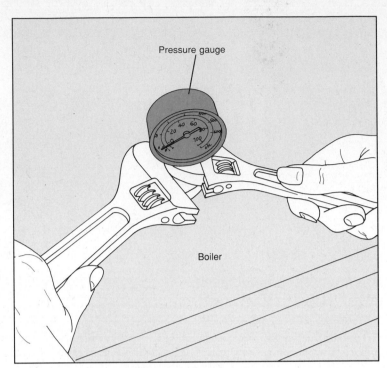

1 Draining the system. Turn off power to the boiler at the main service panel *(page 134)* and unit disconnect switch *(page 135)*. Turn off the water to the boiler, attach a hose to the drain cock and open a bleed valve *(page 42)*. Open the drain cock and let the system drain *(above)*; this can be a long process. An unpleasant odor is normal. When water no longer flows, open the pressure reducing valve; also open a ground-floor bleed valve to make sure there is no water in the system. Close the drain cock.

2 Accessing the boiler. With the boiler drained, remove the pressure gauge *(above)* or safety valve—whichever is boiler-mounted. Hold the boiler fitting with one wrench, and use another wrench to loosen the pressure gauge or safety valve from the fitting *(above)*. Unscrew the gauge or valve by hand.

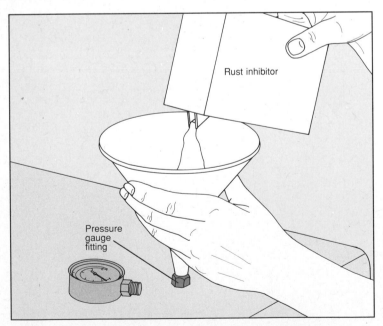

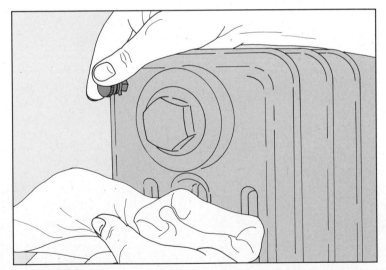

3 Adding rust inhibitor. Buy rust inhibitor, available from a heating supplies dealer. Insert a funnel into the pressure gauge or safety valve opening and, following the label instructions, pour the recommended amount of rust inhibitor for your system into the boiler *(above)*. Reinstall the pressure gauge or safety relief valve.

4 Refilling the system. Close the bleed valves that you opened in step 1, then slowly open the water shutoff valve. **Caution:** If any part of the system is still hot and cold water enters too quickly, the boiler could be damaged. Turn on the water supply to the boiler at the shutoff valve. When the boiler pressure reaches about 5 psi, bleed every radiator or convector in the house, starting on the ground floor. When boiler pressure stabilizes (on most systems, at 15 to 20 psi), restore power to the boiler. Allow the heated water to circulate for several hours, until the radiators or convectors are warm, to purge the water of dissolved air. Then bleed each radiator or convector again.

REPAIRING A LEAKY INLET VALVE

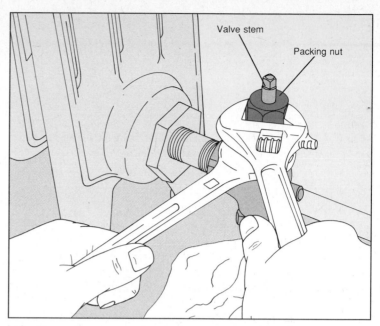

1 Unscrewing the handle. If a radiator or convector inlet valve leaks, try tightening the packing nut one-quarter turn. (Do not over-tighten.) If the valve still leaks, use a screwdriver to remove the handle screw, located in the middle of the valve stem *(above)*. Pull the handle straight off the stem.

2 Removing the packing nut. Wrap rags around the valve area; when you loosen the packing nut, the leaking may increase. Use two wrenches; one to hold the valve fitting steady, and the other to loosen the packing nut from the valve stem, turning counterclockwise *(above)*. Unscrew the packing nut by hand and slip it off the valve stem. If excessive water leaks, reinstall the nut and drain the system *(page 44)*.

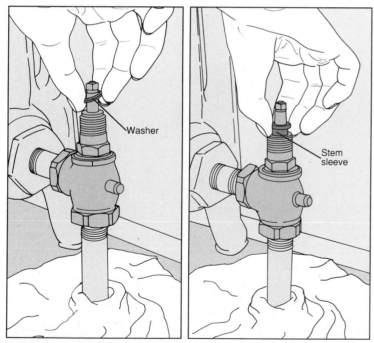

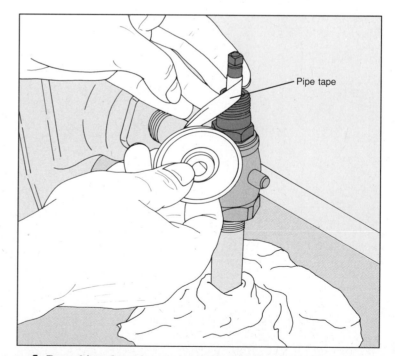

3 Dismantling the valve. With the packing nut removed, pull the washer, if any, off the stem *(above, left)*. Use the blade of a flat-tipped screwdriver to pry up the stem sleeve, if any. Once it is loosened, slip it off the stem by hand *(above, right)*.

4 Repacking the valve. Leaving the old packing in place, add pipe tape *(above)* or new packing string, available from a plumbing supplies dealer. Wrap the tape or string several times around the old packing, stretching it and pressing it down as you go. Replace the stem sleeve and washer, and reinstall the packing nut. Use a wrench to tighten the packing nut until it stops leaking. Screw on the valve handle.

REPLACING A DIAPHRAGM-TYPE EXPANSION TANK

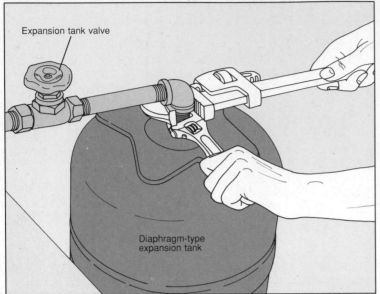

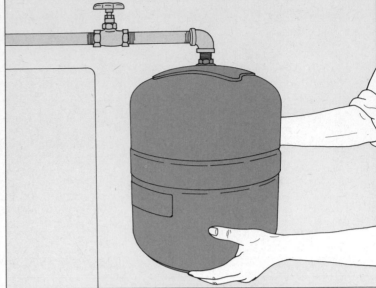

Freeing the expansion tank. Turn off power to the boiler at the main service panel *(page 134)* and unit disconnect switch *(page 135)*, and let the boiler cool. Close the expansion tank valve. Use a pipe wrench to grip the fitting connecting the expansion tank to the pipe *(above, left)*. Use another wrench to loosen the tank just until water starts to drip; wait until the water stops dripping. **Caution:** the tank is full of water, and will be heavy. Have a helper hold it for safety. Twist the tank off the pipe fitting by hand *(above, right)*. Purchase a new expansion tank from a heating supplies dealer. Wrap plumber's tape around the threads. Install the tank, reversing these instructions. Open the expansion tank valve slowly, then restore power to the boiler.

SERVICING AN OLD-STYLE EXPANSION TANK

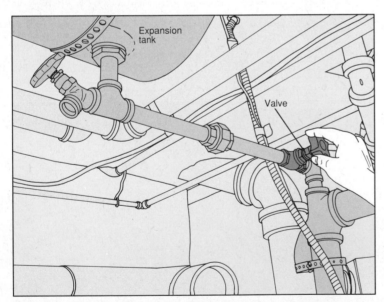

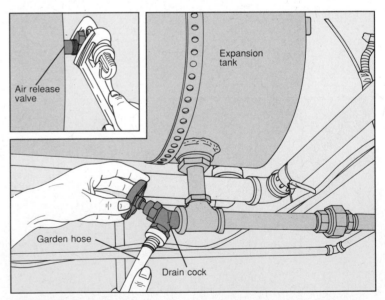

1 **Shutting off water to the tank.** If the pressure gauge reading rises to 30 psi or above and the safety valve spouts water, the expansion tank may be waterlogged and must be recharged. Turn off power to the boiler at the main service panel *(page 134)* and unit disconnect switch *(page 135)*, and allow the boiler to cool. Close the expansion tank valve *(above)*.

2 **Draining and refilling the tank.** Connect a garden hose to the drain cock on the bottom of the expansion tank. Open the drain cock; then use a pipe wrench to open the air release valve on the side of the tank *(inset)*. When the tank has drained completely, tighten the air release valve, close the drain cock and partially open the tank's shutoff valve. You will hear water slowly entering the tank; after the sound stops, open the valve completely. When the pressure gauge reads about 15 psi, restore power to the boiler. When the radiators or convectors are warm, bleed all of them *(page 42)*, beginning on the top floor.

AQUASTATS

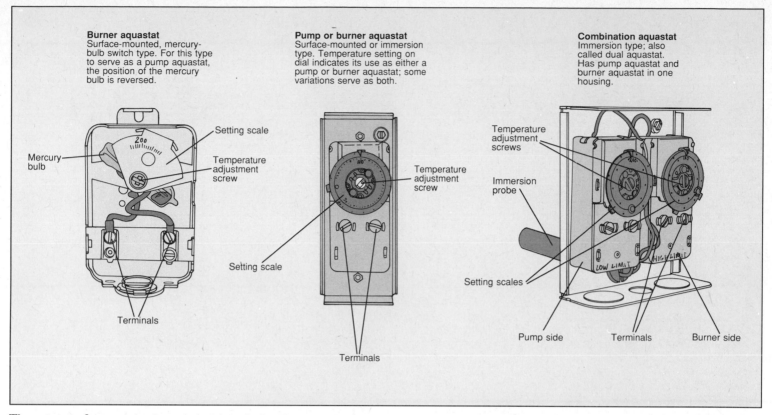

Burner aquastat
Surface-mounted, mercury-bulb switch type. For this type to serve as a pump aquastat, the position of the mercury bulb is reversed.

Mercury bulb

Setting scale

Temperature adjustment screw

Terminals

Pump or burner aquastat
Surface-mounted or immersion type. Temperature setting on dial indicates its use as either a pump or burner aquastat; some variations serve as both.

Temperature adjustment screw

Setting scale

Terminals

Combination aquastat
Immersion type; also called dual aquastat. Has pump aquastat and burner aquastat in one housing.

Temperature adjustment screws

Immersion probe

Setting scales

LOW LIMIT

HIGH LIMIT

Pump side   Terminals   Burner side

Three types of aquastats. An aquastat is a device that senses the temperature of water in the system, and responds by turning the burner or circulator pump on or off, as appropriate. All systems have at least one aquastat; some have as many as three. An aquastat may be surface-mounted—located on a pipe; or the immersion type—with a probe inserted into in a well in the boiler. All boilers have a burner aquastat *(above, left)* that acts as a safety switch, turning off the burner if the water temperature rises above a safe level, usually around 200°F. A second aquastat, the pump aquastat *(above, center)*, turns on the circulator pump when the temperature of the water is high enough to heat the house (usually around 110°F). When the water cools down, it turns off the pump. On some systems, a combination aquastat serves as both pump and burner aquastat *(above, right)*.

TESTING AN AQUASTAT

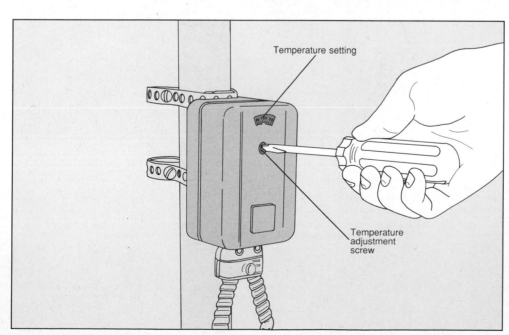

Temperature setting

Temperature adjustment screw

Testing the aquastat. Raise the thermostat to 80°F; the burner should go on. Then locate the temperature adjustment screw on the cover of the aquastat. On a burner aquastat, note the temperature setting, then use a screwdriver to lower the setting below 100°F *(left)*; the burner should turn off. Raise the setting to its original position; after a slight delay (up to 10 minutes on an oil burner with a stack-mounted relay), the burner should turn on. If the aquastat fails either test, it is faulty; replace it *(page 48)*.

On a pump aquastat, note the temperature setting, then use a screwdriver to raise the setting above 100°F; the pump should turn on. Then lower the setting below 100°F; the pump should turn off. If the aquastat fails either test, it is faulty; replace it *(page 48)*. Reset the thermostat.

REPLACING AN AQUASTAT

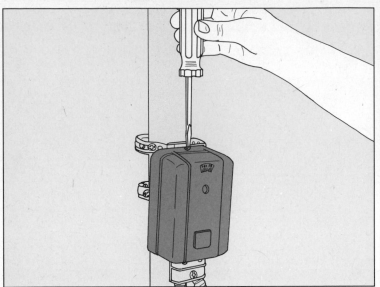

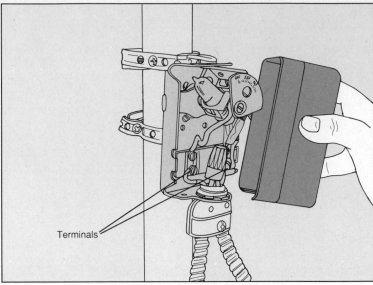

Terminals

1 **Removing the cover.** Turn off power to the boiler at the main service panel *(page 134)* and unit disconnect switch *(page 135)*, and let the boiler cool. On either a surface-mounted aquastat, as shown, or an immersion type, use a screwdriver to loosen the cover retaining screw *(above, left)*. Pull off the cover to expose the terminal screw connections *(above, right)*.

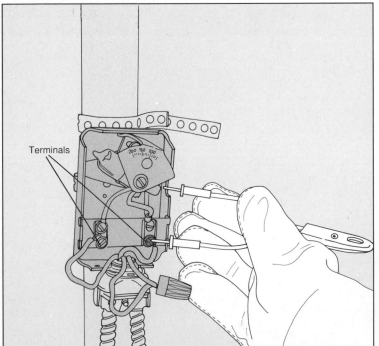

Terminals

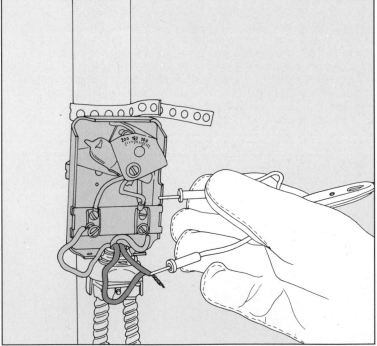

2 **Testing that the power is off.** As a safety precaution, use a voltage tester *(page 132)* to confirm that power is indeed off before disconnecting any aquastat wires. Working with one hand only, touch one tester probe to each of the aquastat line-voltage terminal screws. Then test between each terminal screw and the grounded aquastat box *(above, left)*. Next, unscrew the wire cap connecting the white line-voltage wires and touch one tester probe to the bare wire ends and the other probe to one terminal screw, then the other. Finally, test between the bare wire ends and the grounded aquastat box *(above, right)*. The tester should not glow in any test. If it does, power to the boiler is still on; turn off power to the proper circuit and test again to make sure power is off. If the aquastat is surface-mounted, go to step 3; if it is an immersion type, go to step 4.

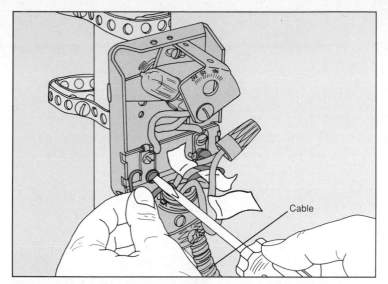

Cable

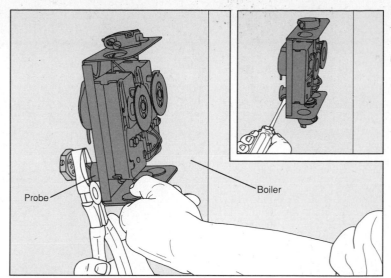

Probe

Boiler

3 **Replacing a surface-mounted aquastat.** With power to the boiler turned off, label the wires inside the aquastat box for correct reconnection. Unscrew the terminals *(above)* and remove the wire caps. Remove the fastener holding the line-voltage cable in place. Unscrew the aquastat from the metal straps holding it to the pipe. Connect an exact replacement aquastat—set on the same setting as the old one—to the pipe with the metal straps. Reconnect the wires, screw on the aquastat cover and restore power to the boiler. Test the new aquastat *(page 47)* to make sure it works properly.

4 **Replacing an immersion aquastat.** With power to the boiler turned off, label the wires inside the aquastat box for correct reconnection. Unscrew the terminals and remove the wire caps. Remove the fastener holding the line-voltage cable in place. Loosen the screws holding the aquastat to the well *(inset)*, and slide the aquastat probe out of the well; use pliers gently to twist the probe if necessary *(above)*. Slide an exact replacement aquastat—set on the same setting as the old one—into place in the well. Reconnect the wires, screw on the aquastat cover and restore power to the boiler. Test the new aquastat *(page 47)* to make sure it works properly.

MAINTAINING A CIRCULATOR PUMP

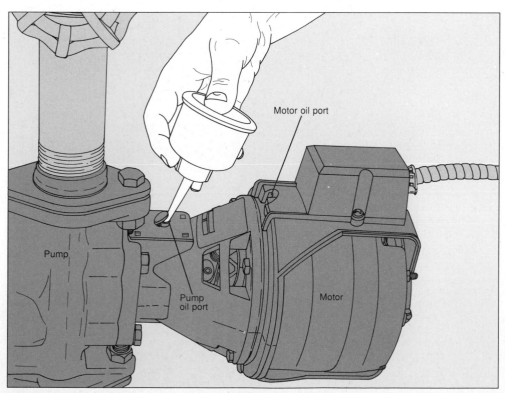

Motor oil port

Pump

Pump oil port

Motor

Lubricating the circulator pump and motor. Once a year—or according to the manufacturer's directions on the plate affixed to the pump—lubricate the pump and motor. (The plate may be covered with dirt or grease; clean it with solvent.) Locate the oil ports, usually three, on top of the circulator pump and motor; use the blade of a flat-tipped screwdriver to pry up the port caps. Squirt SAE-30 non-detergent motor oil into the ports *(left)*. Newer models, such as the one shown, generally require about 30 drops of oil in the pump oil port, and about 15 drops in each motor oil port; older models usually require fewer drops, applied more frequently. Do not force oil into the ports.

TESTING AND REPLACING A CIRCULATOR MOTOR AND COUPLER

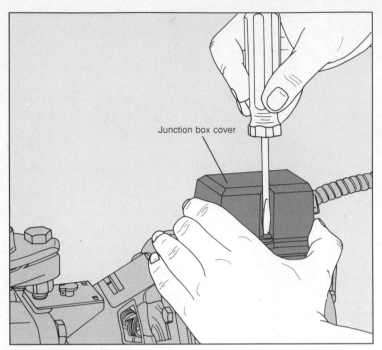

Junction box cover

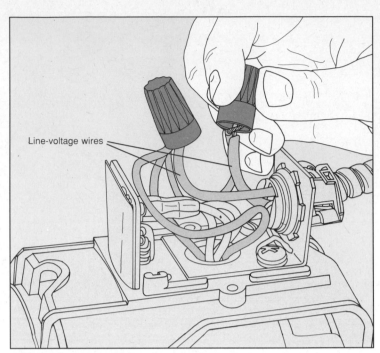

Line-voltage wires

1 **Removing the junction box cover.** Turn off power to the boiler at the main service panel *(page 134)* and unit disconnect switch *(page 135)*. If your system has a pump aquastat, test at the aquastat to make sure power is indeed off *(page 48, step 2)*. Unscrew the screws holding the junction box cover to the top of the circulator motor *(above)* and take off the cover.

2 **Unscrewing the wire caps.** Locate the wire caps connecting the circulator motor to the line-voltage wires. Taking care not to touch any exposed wire ends, unscrew the wire caps.

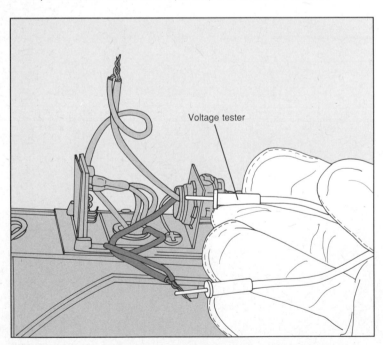

Voltage tester

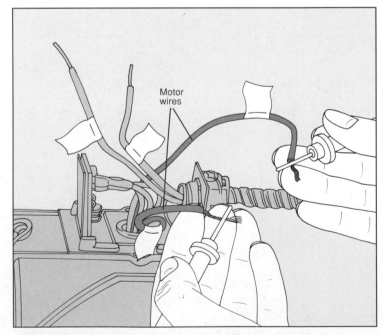

Motor wires

3 **Testing to make sure power is off.** As a safety precaution, use a voltage tester *(page 132)* to confirm that power to the circulator motor is indeed off. Touch one tester probe to the grounded junction box on the motor and the other probe to each exposed wire pair, in turn *(above)*. Then test between the two wire ends. The voltage tester should not glow in any test. If it does, locate the correct circuit, turn it off, then test again to make sure power is off.

4 **Testing the motor windings.** Set a multitester to the RX1K setting to test continuity *(page 136)*. Locate the two motor wires that were connected to the line-voltage wires with wire caps. Label the wires for correct reconnection *(page 135)* and separate the wire ends. Touch a multitester probe to each motor wire *(above)*. If the tester shows continuity, the motor is OK; if there is no continuity, replace the coupler and motor *(page 51)* or take the motor for professional service.

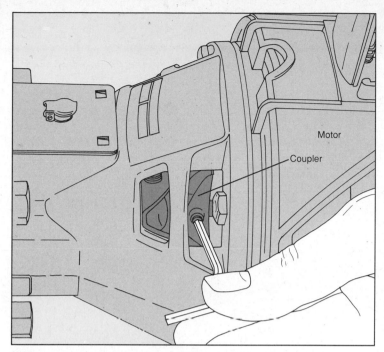

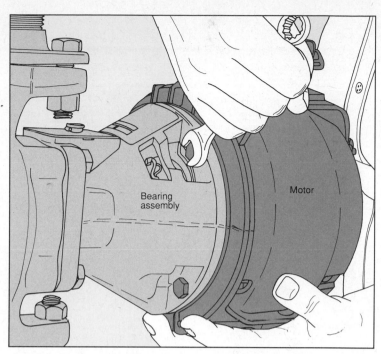

5 **Releasing the coupler from the motor shaft.** If the pump or motor jams, the coupler—a spring-loaded connector between the pump shaft and motor shaft—will come apart, preventing the motor from overheating. Reach into the housing and rotate the motor with the end of a hex wrench until you can see the coupler's motor-shaft setscrew. Use the hex wrench to loosen the setscrew holding the coupler to the motor shaft *(above)*.

6 **Unbolting the motor.** Use a wrench to remove the bolts holding the motor to the bearing assembly *(above)*. Pull the motor free of the bearing assembly and pry the coupler off the motor shaft with a screwdriver if it sticks. If the coupler is broken, replace it *(step 7)*; if you suspect that the motor is faulty, test it *(step 6)*.

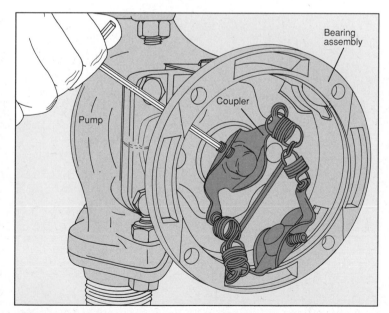

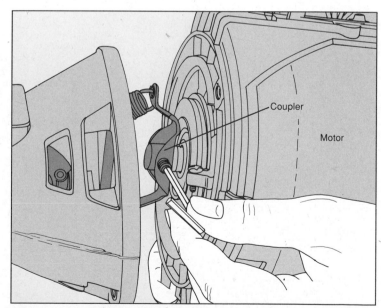

7 **Replacing the coupler.** Use a hex wrench to loosen the setscrew holding the coupler to the pump shaft *(above)*. Once the coupler is loosened, work it off the shaft by twisting it; if necessary, pry it off with a screwdriver. Replace the entire coupler with an identical model; do not attempt to replace the springs. Line up the new coupler setscrew with the setscrew hole on the pump shaft, then use a hex wrench to tighten the setscrew.

8 **Replacing the motor.** Hold the motor in one hand and slide the coupler over the motor shaft. Line up the setscrew with the setscrew hole and use a hex wrench to tighten the setscrew *(above)*. Reinstall the bolts securing the motor to the bearing assembly in a crisscross sequence: first, the bolt in the 12 o'clock position, then 6 o'clock, then 3 o'clock, then 9 o'clock. Reconnect the wiring and screw on the junction box cover. Restore power to the boiler and allow the system to warm up to 110°F before judging whether or not the new motor works.

OIL BURNERS

The oil burner is a team of small machines whose job is to produce and sustain a flame. The flame warms air or water, which circulates through the home to provide heat. Whether used to fire a hot-water, forced-air or steam heating system, the burner operates the same way.

The pump supplies oil to the system, pressurizing it to about 100 pounds per square inch (psi). The oil is forced through a tiny opening in the nozzle, creating a finely-atomized mist that burns quickly and completely when mixed with air. In order to ignite this mixture, the ignition transformer raises household electrical current from roughly 120 volts to 10,000 volts. The stepped-up voltage runs to a pair of electrodes, which create a high-voltage spark that ignites the oil to start the system. The flame then continues burning on its own in the combustion chamber, fed by air supplied by the blower.

Repairs can vary, depending on your particular burner and system. Many models, for example, have a single-line fuel-supply system, with an aboveground oil tank. This system recycles excess oil within the pump; if air enters the oil line, the pump must be primed *(page 57)*. A double-line system, however, uses one line to draw oil to the pump, and a second line to return excess oil to the tank, which is usually underground. This system is designed to be self-priming.

To stop the flow of oil if the burner flame goes out and the thermostat is still calling for heat, an oil burner uses one of two sensing devices: A stack heat sensor shuts off a safety switch

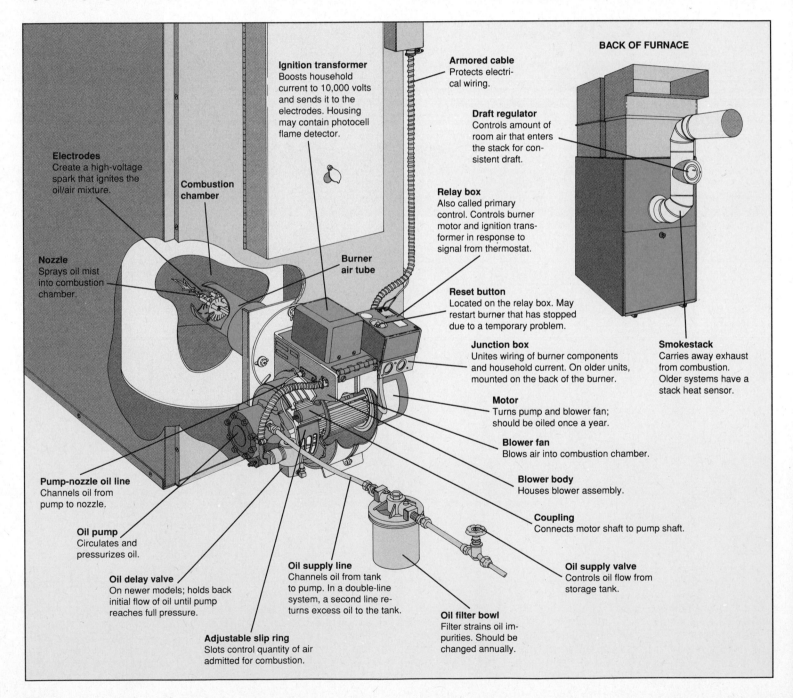

BACK OF FURNACE

Ignition transformer
Boosts household current to 10,000 volts and sends it to the electrodes. Housing may contain photocell flame detector.

Armored cable
Protects electrical wiring.

Draft regulator
Controls amount of room air that enters the stack for consistent draft.

Electrodes
Create a high-voltage spark that ignites the oil/air mixture.

Combustion chamber

Relay box
Also called primary control. Controls burner motor and ignition transformer in response to signal from thermostat.

Nozzle
Sprays oil mist into combustion chamber.

Burner air tube

Reset button
Located on the relay box. May restart burner that has stopped due to a temporary problem.

Junction box
Unites wiring of burner components and household current. On older units, mounted on the back of the burner.

Smokestack
Carries away exhaust from combustion. Older systems have a stack heat sensor.

Motor
Turns pump and blower fan; should be oiled once a year.

Blower fan
Blows air into combustion chamber.

Pump-nozzle oil line
Channels oil from pump to nozzle.

Blower body
Houses blower assembly.

Oil pump
Circulates and pressurizes oil.

Coupling
Connects motor shaft to pump shaft.

Oil delay valve
On newer models; holds back initial flow of oil until pump reaches full pressure.

Oil supply line
Channels oil from tank to pump. In a double-line system, a second line returns excess oil to the tank.

Oil supply valve
Controls oil flow from storage tank.

Oil filter bowl
Filter strains oil impurities. Should be changed annually.

Adjustable slip ring
Slots control quantity of air admitted for combustion.

when no heat is detected after the burner has been running; a photoelectric cell reacts when the combustion chamber goes dark. Both devices are equipped with a reset button that can sometimes get the system operating again. **Caution:** Never press the reset button more than twice; the unburned oil pumped into the combustion chamber can accumulate and may explode or overheat when ignited.

Periodic maintenance and efficiency tests can reduce the need for repairs, as well as lowering your fuel bill. In addition to cleaning the air intake gate and oiling the motor *(page 54)*, check the fuel tank gauge regularly to avoid running out of fuel. To conduct the tests in this chapter, you will need special tools available from a heating supplies dealer: a stack ther-

mometer, a draft-measuring device such as an inclined manometer, a pump pressure gauge and an electrode gauge. Other tools, and the proper way to use them, are presented in Tools & Techniques *(page 132)*.

Follow all safety instructions in this chapter and in the Emergency Guide *(page 8)*. Before beginning most repairs or inspections, turn off power to the burner at the main service panel or flip off the unit disconnect switch.

If your oil-fired heating system is not working properly, the cause may be the oil burner, or it may be another unit in the heating system. Also consult the Troubleshooting Guides for System Controls *(page 16)*, and Water Distribution *(page 39)* or Air Distribution *(page 24)*.

TROUBLESHOOTING GUIDE continued ▶

SYMPTOM	POSSIBLE CAUSE	PROCEDURE
No heat or insufficient heat	No fuel	Check fuel tank gauge; have tank refilled if necessary
	No power to circuit	Replace fuse or reset circuit breaker *(p. 134)* □○
	Relay or stack heat sensor tripped	Press reset button no more than twice; if the system doesn't start, call for service
	Oil leaking at pump fittings	Tighten fittings; if no improvement, replace pump *(p. 61)* ◨●
	Electrodes faulty	Check electrodes; adjust *(p. 67)* ◨●▲ if necessary
	Nozzle faulty	Replace nozzle *(p. 67)* ◨○
	Photocell dirty or faulty	Clean photocell *(p. 55)* □○; replace if necessary
	Stack heat sensor dirty	Clean soot buildup out of stack heat sensor *(p. 55)* □○
	Coupling broken	Replace coupling *(p. 65)* ◨●
	Pump seized	Test pump pressure *(p. 60)* □○▲; take pump for service or replace *(p. 61)* ◨● if necessary
	Motor seized or overloaded	Press reset button; if motor is silent or hums but does not turn, check and replace motor *(p. 63)* ◨●
	Pump pressure too low	Test pump pressure *(p. 60)* □○▲; take pump for service or replace *(p. 61)* ◨● if necessary
	Oil filter dirty	Replace oil filter *(p. 57)* □○
	Oil strainer dirty	Clean strainer *(p. 59)* □○; replace if necessary
Intermittent heat	Fuel supply low	Check fuel tank gauge; have tank refilled if necessary
	Oil filter dirty	Replace oil filter *(p. 57)* □○
	Oil strainer dirty	Clean strainer *(p. 59)* □○; replace if necessary
	Pump pressure too high	Test pump pressure *(p. 60)* □○▲; take pump for service or replace *(p. 61)* ◨● if necessary
	Motor overloaded	Press reset button; if motor is silent or hums but does not turn, check and replace motor *(p. 63)* ◨●
	Pump seized	Test pump pressure *(p. 60)* □○▲; take pump for service or replace *(p. 61)* ◨● if necessary
	Nozzle faulty	Replace nozzle *(p. 67)* ◨○
	Photocell dirty or faulty	Clean or replace photocell *(p. 55)* □○
	Stack heat sensor dirty	Clean soot buildup out of stack heat sensor *(p. 55)* □○

DEGREE OF DIFFICULTY: □ **Easy** ◨ **Moderate** ■ **Complex**
ESTIMATED TIME: ○ **Less than 1 hour** ● **1 to 3 hours** ● **Over 3 hours** ▲ **Special tool required**

TROUBLESHOOTING GUIDE (continued)

SYMPTOM	POSSIBLE CAUSE	PROCEDURE
High fuel consumption	Nozzle faulty	Replace nozzle (p. 67) ■○
	Pump pressure too low or too high	Test pump pressure (p. 60) □○▲ ; take pump for service or replace (p. 61) ■◐ if necessary
Burner system noisy	Motor bearings dry	Lubricate motor (p. 54) □○
	Coupling misaligned, loose or broken	Adjust, tighten or replace coupling (p. 65) ■◐
	Pump worn out	Test pump pressure (p. 60) □○▲ ; take pump for service or replace (p. 61) ■◐ if necessary
	Draft regulator misadjusted	Test draft and adjust counterweight (p. 56) ■◐▲
	Incorrect oil/air mix causing flame roar	Test stack temperature (p. 56) □○▲ ; call for service
Diesel fuel odor from burner system	Draft regulator wide open	Adjust counterweight (p. 56) □○
	Stack or heat exchanger blocked with soot	Test stack temperature (p. 56) □○▲ ; call for service to have stack or heat exchanger cleaned
	Pump leaking	Replace pump (p. 61) ■◐
	Nozzle faulty	Replace nozzle (p. 67) ■○
	Electrodes faulty	Check electrodes (p. 67) ■◐▲ ; adjust if necessary
	Stack damaged	Call for service
Electrical insulation odor from burner system	Ignition transformer overheating	Test transformer (p. 65) ■◐ ; replace if necessary
	Motor shorted	Replace motor (p. 63) ■◐

DEGREE OF DIFFICULTY: □ Easy ■ Moderate ■ Complex
ESTIMATED TIME: ○ Less than 1 hour ◐ 1 to 3 hours ● Over 3 hours
▲ Special tool required

CLEANING AND LUBRICATING THE BURNER

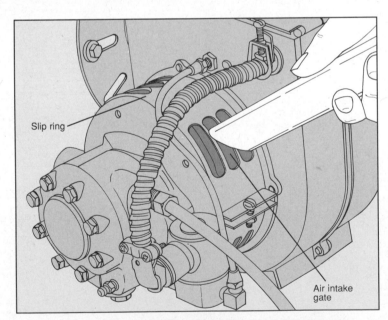

Vacuuming the air intake gate and blower. Turn off power to the burner at the main service panel *(page 134)* or unit disconnect switch *(page 135)*. Locate the air intake gate, usually at the left side of the burner assembly when you face the furnace or boiler; it is covered by a slotted slip ring. As air enters through the slots, dust, animal hair or dryer lint can accumulate. Using a crevice tool, vacuum the air intake gate *(above)*. If necessary, mark the slip ring position *(page 62)* and detach it to remove heavy buildup. Reinstall the slip ring before turning on power to the burner.

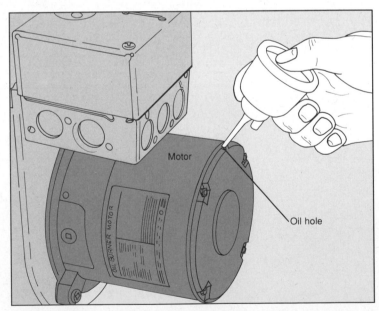

Oiling the motor. Turn off power to the burner at the main service panel *(page 134)* or unit disconnect switch *(page 135)*. Locate the motor, usually at the right side of the burner assembly when you face the furnace or boiler. The motor has one or more oil holes at the top. Using an oil can, insert a drop of high-grade SAE 30 machine oil into each hole *(above)*. Oil sparingly, a total of four drops once per year; too much oil, or oiling too often, can damage the internal starting switch. Turn on power to the burner.

SERVICING THE PHOTOCELL FLAME DETECTOR

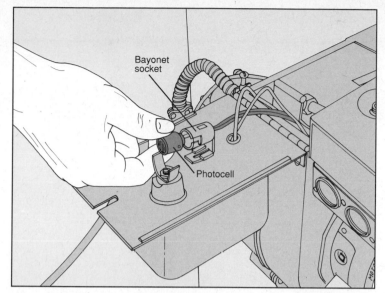

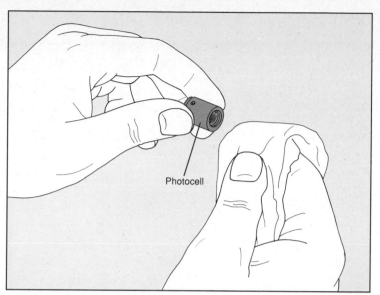

1 **Removing the photocell.** If your burner has no photocell, remove and clean the stack heat sensor *(below)*. On a unit with a photocell, turn off power to the burner at the main service panel *(page 134)* or unit disconnect switch *(page 135)*. Remove the screw from the transformer's hinged hatch and open the hatch. Remove a bayonet-style photocell by pushing it in, then twisting it out of its socket *(above)*; pull out a plug-in photocell.

2 **Cleaning the photocell.** Use a clean, moist cloth to wipe soot buildup off the photocell "eye" *(above)*. Reinstall the photocell in its socket, close the transformer hatch and reinstall the screw. Turn on power to the burner. If the photocell is still faulty, remove it again and replace it with an identical photocell.

SERVICING THE STACK HEAT SENSOR

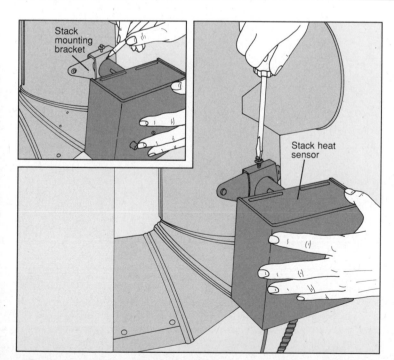

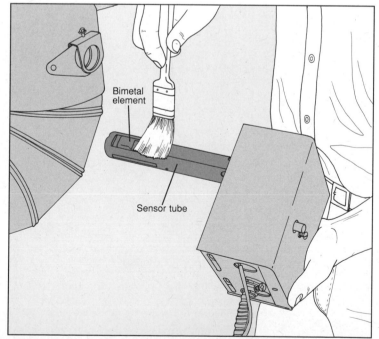

1 **Removing the stack heat sensor.** Turn off power to the burner at the main service panel *(page 134)* or unit disconnect switch *(page 135)*. Use a felt-tip pen to mark the tube of the heat sensor where it meets the stack mounting bracket *(inset)*. Loosen the lock screw on the mounting bracket *(above)* and pull the stack heat sensor out of the stack.

2 **Cleaning the sensor tube and bimetal element.** Use a soft paintbrush *(above)* to sweep away soot that has collected on the heat-sensor bimetal element and sensor tube. Replace the stack heat sensor to the depth indicated by the pen mark and tighten the mounting bracket screw. Turn on power to the burner.

TESTING THE STACK TEMPERATURE

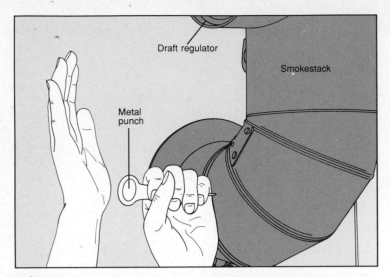

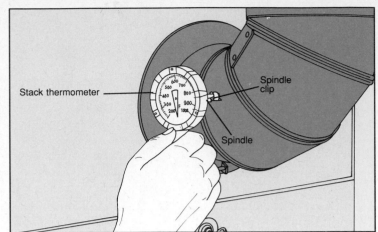

1 **Punching a hole in the stack.** Locate the smokestack, a galvanized pipe roughly 6 inches in diameter that emerges from the furnace. If a hole was previously drilled in the stack for testing, remove the sheet metal screw blocking it, then go to step 2. If there is no hole, make a hole about a foot from the furnace, at least 6 inches below the draft regulator. Use a 1/4-inch punch, striking the punch with the heel of your hand *(above)*. Set up an indoor thermometer in the furnace room, several feet from the furnace, to measure room temperature.

2 **Reading the stack thermometer.** Set the thermostat 5°F above room temperature and allow 20 minutes for the burner to warm up. Insert the spindle of a stack thermometer into the hole *(above)*, and secure it with the spindle clip. Leave the thermometer in place until the temperature has stabilized, then subtract the furnace room temperature from the stack temperature. A result between 400°F and 750°F is normal; block the stack hole with a sheet metal screw. If the result is below 400°F, the burner needs adjustment or the nozzle is faulty; replace the nozzle *(page 67)* or call for service. A result above 750°F may indicate excessive draft; test and adjust the draft regulator *(below)*.

TESTING AND SETTING THE DRAFT

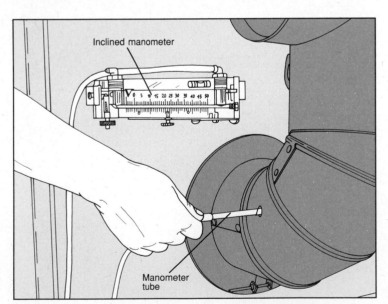

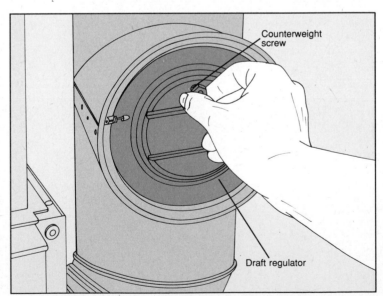

1 **Testing draft in the stack.** Make a hole in the stack *(step 1, above)*. Set the thermostat 5°F above room temperature to turn on the blower. Using an inclined manometer *(page 132)*, measure air movement through the stack. Mount the manometer according to the instructions accompanying it, making sure its spirit level is horizontal. Insert the manometer tube halfway into the stack *(above)*, then read the gauge. A reading between -.02 and -.05 is within the normal range. (The precise reading for your particular burner may be printed on the burner housing or in your burner manual.) If the reading is not between -.02 and -.05, adjust the draft regulator *(next step)*.

2 **Adjusting the draft regulator.** Turning the counterweight screw allows more or less air to enter the stack. Increase the air flow, or draft, by turning the screw counterclockwise; decrease the draft by turning it clockwise *(above)*. Retest the draft as in step 1, adjusting the screw until the correct draft reading is achieved. Then plug the test hole with a sheet metal screw.

PRIMING THE PUMP

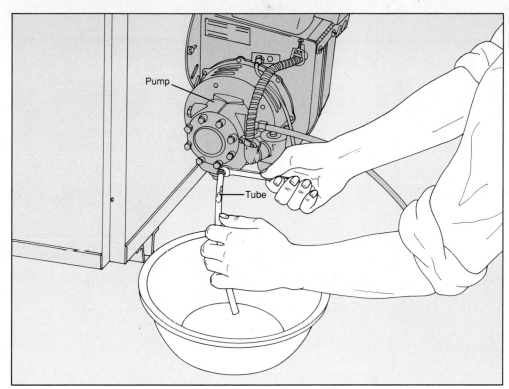

Pump

Tube

Loosening the bleeder nut. Set the thermostat 5°F above room temperature. If you have turned off power to the burner and closed the main oil-supply valve in order to work on the burner, restore the power and open the oil supply valve. If you suspect that the safety control system has shut off the burner, locate the reset button on the relay box and press it no more than twice to reset the relay. Once the burner is running, locate the bleeder nut, a small plug with a nipple in its center, on the pump. Slip a flexible clear-plastic tube over the bleeder nut nipple, and set out a container to catch dripping oil. Using an open-end wrench, slowly loosen the bleeder nut by turning it counterclockwise three-quarters of a turn *(left)*; you will see air bubbles in the oil flow. Bleed the pump until the oil runs smoothly, with no air bubbles. Tighten the bleeder nut and remove the tube. If the relay system shuts off the burner during bleeding, press the reset button once to reset the relay, then continue bleeding.

REPLACING THE OIL FILTER

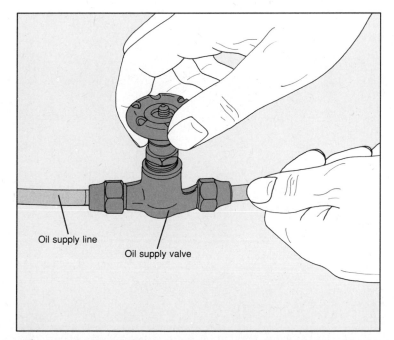

Oil supply line

Oil supply valve

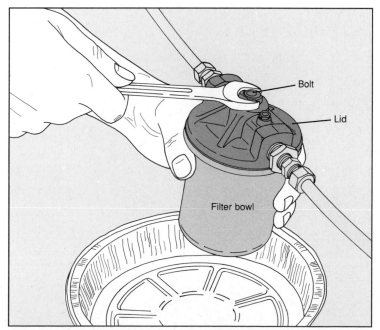

Bolt

Lid

Filter bowl

1 **Closing the oil supply valve.** Turn off power to the burner at the main service panel *(page 134)* or unit disconnect switch *(page 135)*. Locate the oil supply valve on the oil supply line between the burner and the oil tank. Both the oil supply valve and the oil filter may be located at either end of the oil line: next to the oil tank or next to the burner assembly. Close the valve by turning its knob clockwise.

2 **Unfastening the filter bowl lid.** Place a bucket or pan under the oil filter. Use an open-end wrench or screwdriver to unfasten the bolt or screw attaching the filter bowl to its lid *(above)*. If the bolt or screw is particularly stubborn, grip the lid firmly to avoid bending the oil line's copper tubing, and wear a glove to protect your other hand when the bolt jolts loose. As you loosen the bolt, hold the filter bowl to prevent it from spilling oil.

REPLACING THE OIL FILTER (continued)

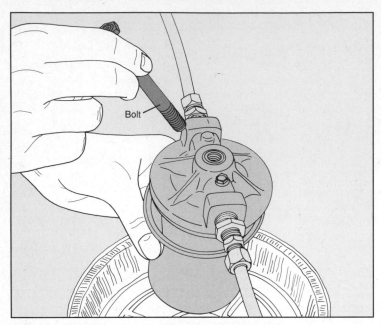

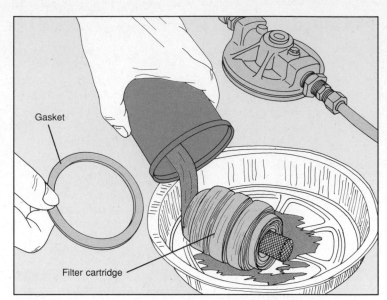

3 **Removing the filter bowl.** Still holding the filter bowl, pull out the long bolt and washer *(above)*. If the filter bowl sticks to its lid, gently twist them apart, or tap the bowl gently with a wooden block. Be careful not to twist the oil line's copper tubing. Pull the bowl straight down off the lid to avoid spilling oil outside the drain pan.

4 **Emptying the filter bowl.** Turn the bowl upside down over the drain pan *(above)*. Oil and a filter cartridge should fall out of the bowl. The gasket around the top of the bowl will usually fall off; if it sticks to the bowl or the lid, pry it loose with a screwdriver. (Except in an emergency, don't worry about keeping the gasket intact; it should be replaced.) Using a clean rag, wipe oil and grit off the inside of the bowl and its lid, then wipe clean the outside of the filter bowl and the surrounding area.

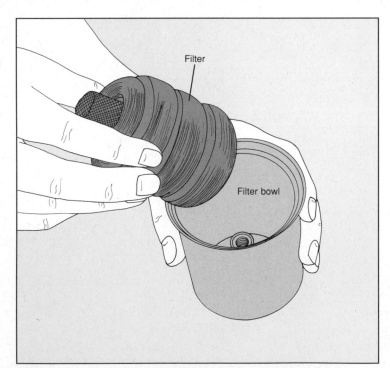

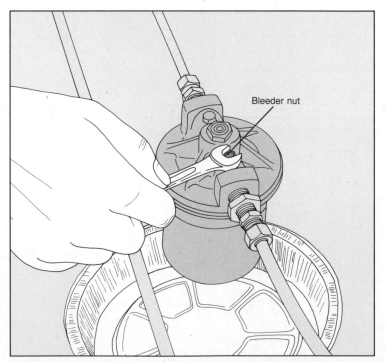

5 **Inserting a new filter cartridge.** Buy exact replacements for the oil filter cartridge and gasket. Push the new filter cartridge into the filter bowl *(above)*. Rub fuel oil on both sides of the new gasket, then fit it onto the lip of the filter bowl. Press the bowl and gasket against the filter bowl lid. Tighten the lid bolt or screw, making sure not to twist the copper tubing of the oil supply line. On a single-line system, prime the pump *(page 57)*. If your system's oil tank is higher than the filter bowl, go to step 6.

6 **Bleeding air from the filter bowl.** Locate the small nut blocking a hole in the lid of the filter bowl; this hole releases air as the filter bowl fills with oil. Use an open-end or adjustable wrench to remove the bleeder nut *(above)*. Open the oil supply valve *(page 57)*; in about 30 seconds the filter bowl should be full of oil. Screw in the bleeder nut and turn on power to the burner.

CLEANING THE PUMP STRAINER

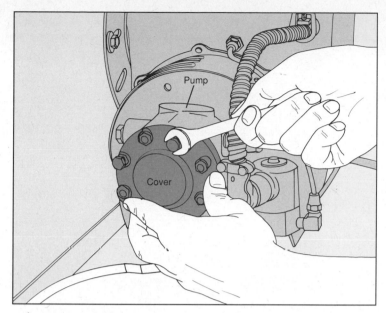

1 Unbolting the pump cover. Turn off power to the burner at the main service panel *(page 134)* or unit disconnect switch *(page 135)*, then close the oil supply valve *(page 57)*. Use an open-end or adjustable wrench to unbolt the pump cover *(above)*, and take it off.

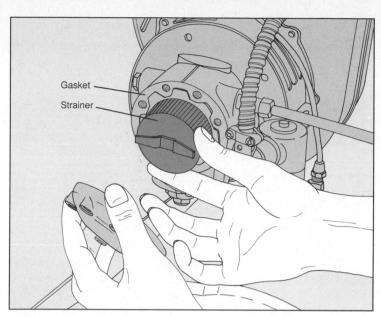

2 Removing the strainer. Locate the strainer, a small, squat cylinder of wire mesh, positioned horizontally inside the pump. (Some older models have a rotary filter, which is removed and cleaned the same way.) Pull the strainer or filter out of the pump *(above)*, and soak it in cleaning solvent for 15 minutes to loosen built-up sludge. Peel off the thin gasket sealing the rim of the pump; buy an identical replacement gasket.

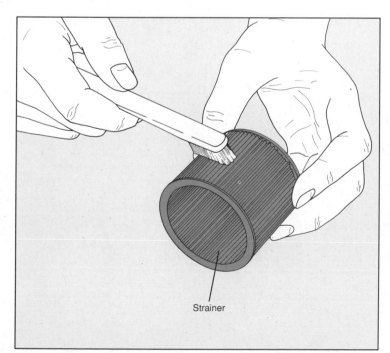

3 Inspecting and cleaning the strainer. Examine the strainer closely for damage or pieces of grit. If the strainer is bent, or the wire mesh is torn, replace it with an identical strainer. Otherwise, use a toothbrush or fine-bristled nylon brush to scrub the strainer mesh *(above)*; a wire brush could damage the mesh. Dislodge all dirt and foreign particles from the mesh, rinse the strainer with solvent, then reinsert it in the pump.

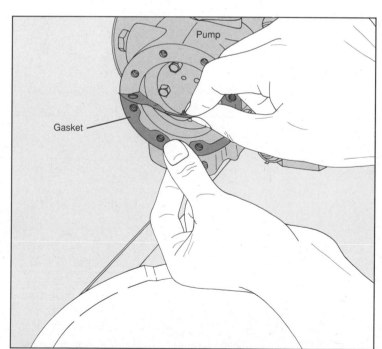

4 Replacing the gasket. Place the adhesive side of the gasket against the pump rim, pressing firmly to ensure a tight bond *(above)*. Replace the pump cover and screw in the bolts by hand. Tighten the bolts in a crisscross sequence: first, the bolt in the 12 o'clock position, then 6 o'clock, then 3 o'clock, then 9 o'clock. Open the oil supply valve *(page 57)* and turn on power to the burner. For a single-line system, prime the pump *(page 57)*.

SERVICING THE RELAY BOX

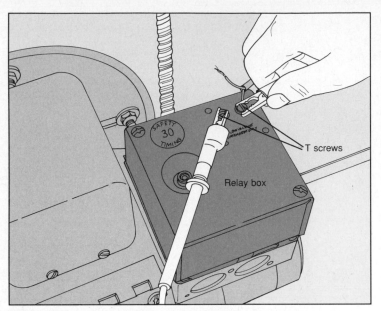

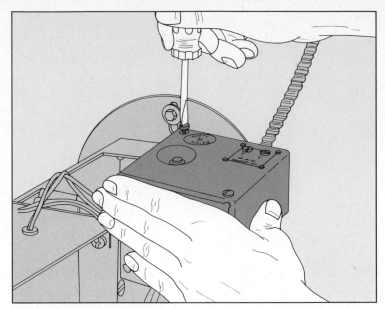

Relay box

T screws

1 **Testing for low voltage.** Turn off power to the burner at the main service panel *(page 134)* or unit disconnect switch *(page 135)*. Disconnect a wire from one of the two thermostat screw connections (labeled "T") on the relay box. On a stack-mounted relay, unscrew the housing and open the box to find the T screws. Set a multitester to measure 24 volts *(page 137)*, then attach a probe with an alligator clip to each T screw. Without touching the relay box or multitester, turn on power to the burner. If the multitester shows a reading of about 24 volts, the relay is OK. Turn off power and reconnect the disconnected wire, then turn on the power. If there is no reading, remove the relay box from the junction box *(step 2)* and check the wiring.

2 **Unscrewing the relay.** With power to the burner turned off, locate the junction box. On newer units, it is located beneath the relay box, usually on top of the motor. On older units, the junction box is generally on the back of the burner; if you have difficulty finding it, trace the armored cable running to the burner. Remove the two screws securing the relay box to the junction box *(above)*. Make sure all connections inside the junction box are tight and that wire caps are correctly fastened *(page 139)*. Tighten loose connections, reattach the junction box and test again *(step 1)*. If the test shows no voltage, turn off the power and replace the old relay box with an identical new one. Transfer the wires one by one from the old box to the new box, then screw the new box in place. Turn on power to the burner.

TESTING PUMP PRESSURE

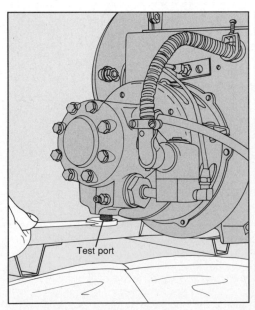

Test port

Return oil line

1 **Accessing the test port.** Turn off power to the burner at the main service panel *(page 134)* or unit disconnect switch *(page 135)*. Place newspaper or a drain pan under the pump to catch dripping oil. Using an open-end wrench, remove the bolt from the test port on the pump *(far left)*. On a two-line system *(near left)*, the test port is sometimes used as a return port for oil; in that case, disconnect the return oil line from the port.

TESTING PUMP PRESSURE (continued)

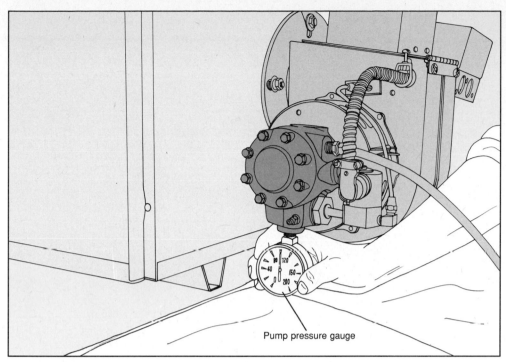

2 **Reading the pump pressure.** Screw a pump pressure gauge—available from a heating supplies dealer—into the threaded opening of the test port *(left)*. Turn on power to the burner. With the pump running, the gauge should show a reading of between 90 and 100 psi (pounds per square inch). If the reading is above or below these levels, turn off the power and remove the gauge. Remove the pump *(below)* and either take it to a repair shop for pressure adjustment, or replace it. If there is no pressure at all, turn off the power, bleed air from the filter bowl *(page 58)*, and test again. If the reading is OK, turn off the power, remove the gauge and reinstall the test port bolt in a single-line system, or the return oil line in a double-line system. If the problem persists, the pump may be seized; check the coupling *(page 65)*.

Pump pressure gauge

REMOVING AND REPLACING THE PUMP

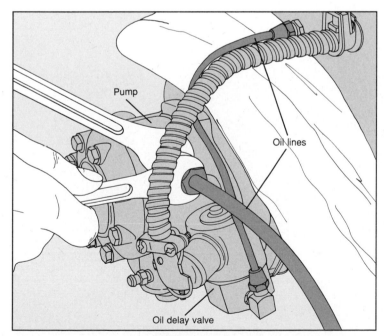

Pump

Oil lines

Oil delay valve

1 **Disconnecting the oil lines.** Turn off power to the burner at the main service panel *(page 134)* or unit disconnect switch *(page 135)*. Close the oil supply valve *(page 57)*. Use a pair of open-end wrenches to loosen the nuts attaching the two oil lines to the pump *(above)*. Take care not to twist the copper tubing of the oil lines. Newer pumps, such as the one shown above, may be equipped with an oil delay valve; disconnect the small oil line from it, then go to step 2 to disconnect its wires. If your pump does not have an oil delay valve, go to step 5.

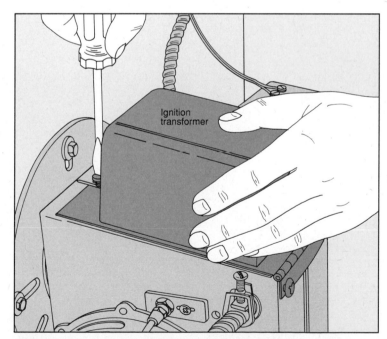

Ignition transformer

2 **Unscrewing the ignition transformer.** Locate the ignition transformer, a black box usually on top of the burner. Remove the screw *(above)*, and tip the transformer on its hinge to expose the wires beneath it. If your model does not have a hinge, remove it and set it on a firm surface. Some older-style transformers are connected to the burner by leads that resemble spark plug wires. If the wires get in the way, label them and unplug them from the transformer. Remove the relay box *(page 60, step 2)* to expose the wiring in the junction box beneath it.

REMOVING AND REPLACING THE PUMP (continued)

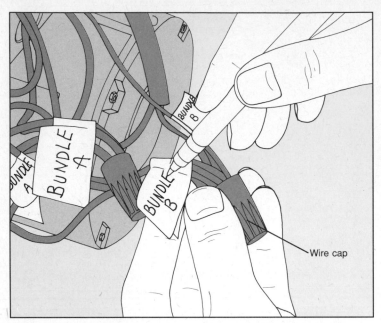

Wire cap

3 **Identifying the wires.** Trace the wires from the oil delay valve into the junction box. The oil delay valve should have two power wires and one ground wire. The two power wires are usually the same color and will be connected to other wires in the junction box, in two bundles held by wire caps. The ground wire should be screwed to the inside of the box. Label each wire in the two bundles *(above)*, designating them "bundle A" and "bundle B."

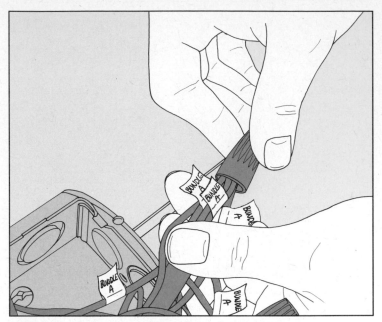

4 **Disconnecting the oil-delay valve wires.** One at a time, twist off the two wire caps by turning them counterclockwise *(above)*, and separate the wires that they connect. Loosen or remove the ground wire screw and disconnect the wire.

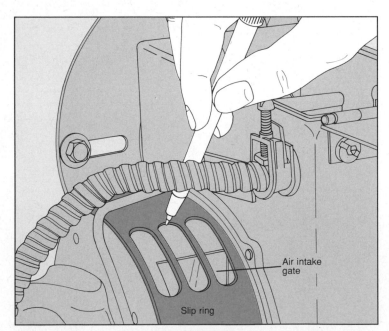

Air intake gate

Slip ring

5 **Marking the slip ring position.** Locate the air intake gate on the side of the burner assembly, usually at the left. The air intake gate, an opening to the blower, is covered with a slotted slip ring. Notice the position of the adjustable slip ring; if this position is altered, serious combustion problems could result. For proper repositioning, use a felt-tip pen to outline the edges of two of the ring's holes on the blower housing *(above)*.

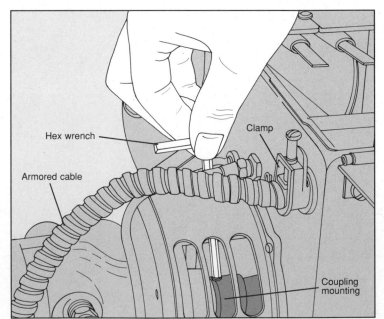

Hex wrench

Clamp

Armored cable

Coupling mounting

6 **Loosening the coupling mounting.** Turn the slip ring until you can see the coupling that connects the pump shaft to the motor shaft—it resembles a short garden hose. If the coupling is held in place by setscrews, insert a hex wrench through a slip ring slot and loosen the coupling mounting at the pump end *(above)*. In some cases, this may require an extra-long hex wrench. If there is no setscrew, go to step 7. Before removing the pump, loosen the clamp screw and pull out the armored cable to free the oil-delay valve wires.

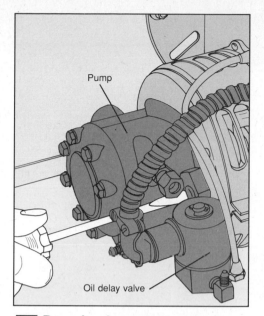

Pump

Oil delay valve

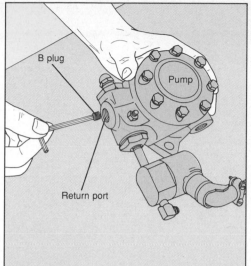

B plug

Pump

Return port

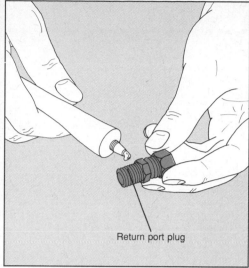

Return port plug

7 **Removing the pump.** Unscrew the pump from the burner housing *(above)*, or use a wrench to remove bolts. Pull the pump, and oil delay valve, if any, off the burner. Have the pump pressure adjusted professionally, or replace the pump with an identical model, keeping the fittings from the old pump.

8 **Preparing the replacement pump.** If you have a double oil-line system, insert a B plug in the new pump before installing it. This plug—which comes with the new pump—works as a bypass, blocking the oil's route back to the pump, and directing it to the tank. **Caution:** do not install a B plug on a single-line system; this can damage the pump. Unscrew the return port plug from the bottom of the pump with an open-end wrench. Use a hex wrench to screw the B plug into the small hole on the inner wall of the return port *(above, left)*. Install the new pump and reconnect the wires, reversing the sequence used to remove the old pump. Install the old pump's fittings, applying liquid sealer to threaded connections *(above, right)*. Open the oil supply valve and turn on power to the burner. If the burner has a single-line system, prime the pump *(page 57)*.

REPLACING THE MOTOR AND BLOWER

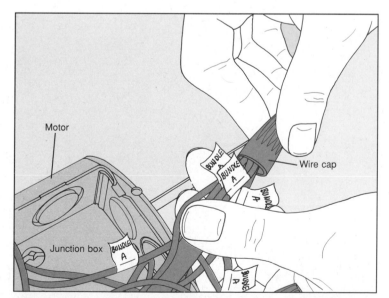

Motor

Wire cap

Junction box

BUNDLE A

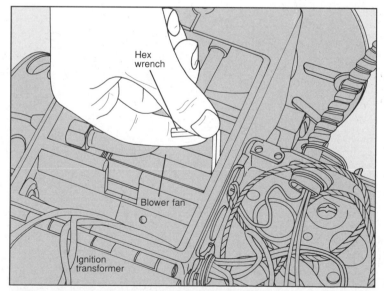

Hex wrench

Blower fan

Ignition transformer

1 **Disconnecting the wiring.** Turn off power to the burner at the main service panel *(page 134)* or unit disconnect switch *(page 135)*. Remove the relay box *(page 60, step 2)* to expose the wires inside the junction box. Locate the two motor wires; they may be encased in a rubber sleeve, as shown. The wires are connected in two bundles held by wire caps. Label each wire in each bundle, designating them "bundle A" and "bundle B." Then twist off the wire caps and unwind the wires *(above)*.

2 **Loosening the coupling mounting.** Unscrew and open the transformer *(page 61)* to access the blower. Locate the coupling that connects the pump and motor shafts—it resembles a short garden hose. If the coupling is not attached with a setscrew, go to step 3. If it is held in place by setscrews, loosen the coupling mounting screw at the motor end by inserting a long hex wrench through the vanes of the blower fan *(above)*. Take care not to damage the vanes.

REPLACING THE MOTOR AND BLOWER (continued)

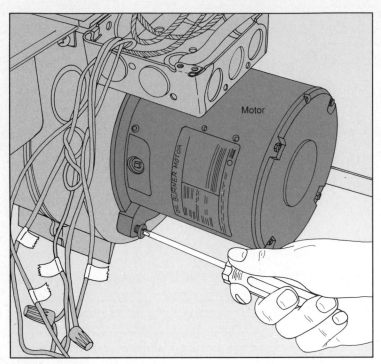

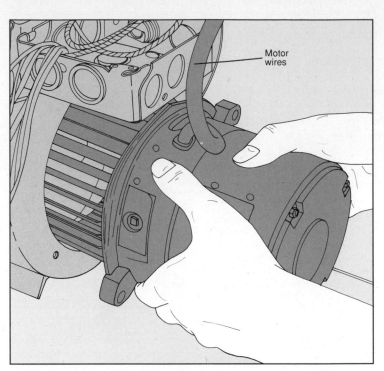

3 **Unfastening the motor.** The motor is generally attached to the side of the burner assembly by two or more bolts or screws. Use an open-end wrench or socket wrench to remove bolts, or a screwdriver to remove screws *(above)*.

4 **Removing the motor and blower.** Pull the motor wires out through the hole in the junction box. Grip the motor firmly with both hands. Taking care not to damage the blower fan, pull the motor and blower assembly straight out *(above)*.

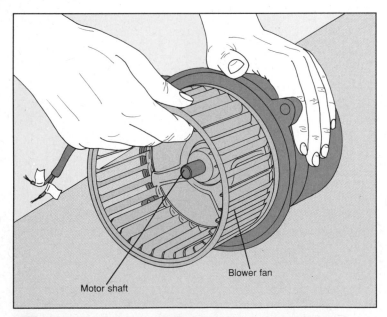

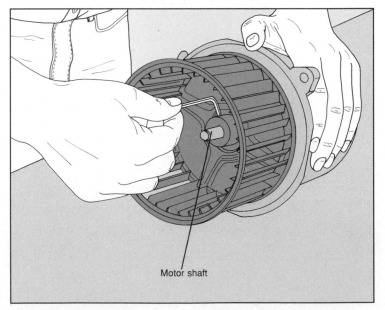

5 **Checking the motor shaft.** Gently rotate the blower fan to turn the motor shaft *(above)*. If it does not turn, the motor is seized; replace it. If the motor shaft turns, reach into the blower cavity on the burner and turn the coupling. If the coupling does not turn, the pump is seized; replace the pump *(page 61)*. If both the pump and motor shafts turn freely, the motor may still be at fault; have it professionally tested.

6 **Changing the blower fan.** Loosen the setscrew attaching the blower fan to the motor shaft *(above)*. Move the blower fan to the new motor shaft. Push the blower fan as far as it will go on the shaft, then pull it back 1/16 inch to prevent friction. Reinstall the motor, reversing the sequence here, and tighten the coupling mounting setscrew if there is one. When reconnecting the motor wires in the junction box, leave no uninsulated wires exposed. Turn on power to the burner.

SERVICING THE COUPLING

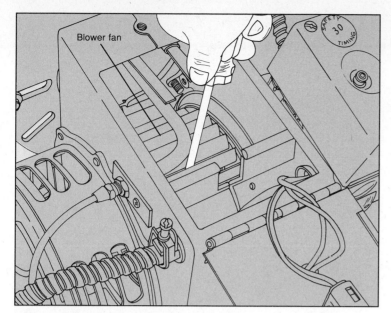

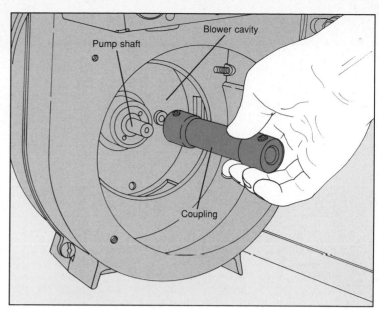

1 **Inspecting the coupling.** Turn off power to the burner at the main service panel *(page 134)* or unit disconnect switch *(page 135)*. Open the transformer to access the blower assembly *(page 61)*. Looking inside the blower fan, locate the coupling; it connects the pump shaft to the motor shaft and resembles a short garden hose. Taking care not to damage the vanes, gently reach through the blower fan with a screwdriver and press on the coupling *(above)*. If it is sturdy, close the transformer box. If the coupling feels loose, spongy or rotted, replace it *(step 2)*.

2 **Checking and replacing the coupling.** If the coupling is held in place by setscrews, loosen it at the pump end *(page 62)*, then loosen it at the motor end and remove the motor *(page 63, steps 1-4)*. Reach into the blower cavity and turn the coupling. If it will not turn, the pump has seized; replace the pump *(page 61)*. If the coupling turns, remove it *(above)*. If the coupling has no setscrews, remove the blower motor *(page 63)* and simply pull off the coupling. Buy an exact replacement coupling and install it by reversing the sequence here. Turn on power to the burner.

SERVICING THE TRANSFORMER

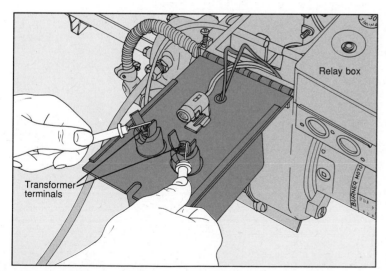

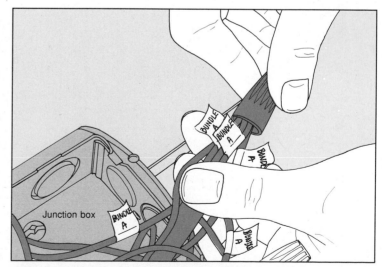

1 **Testing the transformer.** Turn off power to the burner at the main service panel *(page 134)* or unit disconnect switch *(page 135)*. Flip open the transformer *(page 61)*. Set a multitester to the RX10K scale *(page 137)* and touch a multitester probe to each transformer terminal *(above)*. If the multitester does not show a reading, replace the transformer *(steps 2 and 4)*. If the multitester does show a reading, go to step 2.

2 **Disconnecting the primary leads.** With power to the burner turned off, unscrew the relay box from the junction box *(page 60, step 2)*. Trace the two wires leading from the transformer to the junction box. The wires are connected in two bundles with wire caps. Label each wire in the two bundles, designating them "bundle A" and "bundle B." Then twist off the wire caps *(above)* and separate the wires.

SERVICING THE TRANSFORMER (continued)

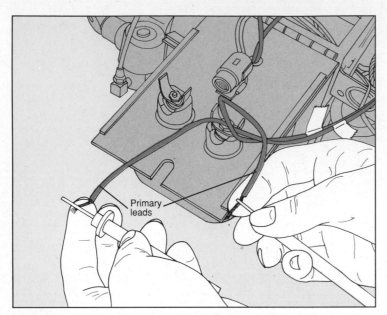

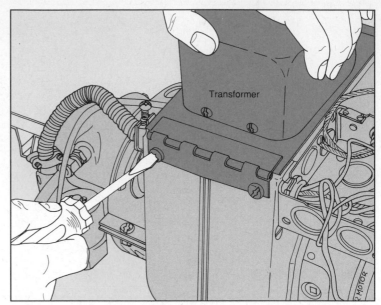

3 **Testing the transformer at the primary leads.** Set a multitester to a resistance scale *(page 137)*, and touch a probe to each primary lead of the transformer. If the multitester shows a reading, the transformer is OK; if there is no reading, replace the transformer.

4 **Unscrewing the transformer.** Flip the transformer back into its normal operating position. Remove the screws securing it to the burner housing *(above)*; or, on some models, unclip the transformer from the housing. Pull the transformer and its wires away from the burner to remove it. Purchase an exact replacement transformer from a heating supplies dealer and install it, reversing the sequence here. Turn on power to the burner.

ACCESSING THE FIRING ASSEMBLY

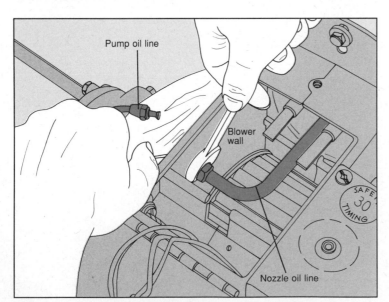

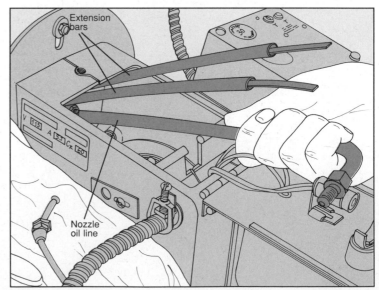

1 **Disconnecting the oil line.** Turn off power to the burner at the main service panel *(page 134)* or unit disconnect switch *(page 135)*. Close the oil supply valve *(page 57)* and open the transformer box *(page 61)*. Locate the nozzle oil line inside the blower body, and the nut connecting it to the pump oil line on the blower wall. Place a rag underneath the pump oil line, and use an open-end wrench to loosen its nut. Then loosen the nut on the nozzle line and pull the pump line free *(above)*.

2 **Removing the firing assembly.** Detach the nozzle oil line from the blower wall. Gripping the oil line firmly, pull the entire firing assembly out of the air tube *(above)*. It may be necessary to turn or twist the assembly slightly as you pull it out. Avoid catching the electrode extension bars on the burner housing, or bumping the electrodes and nozzle. Replace the nozzle or clean and adjust the electrodes as necessary *(page 67)*.

REPLACING THE NOZZLE

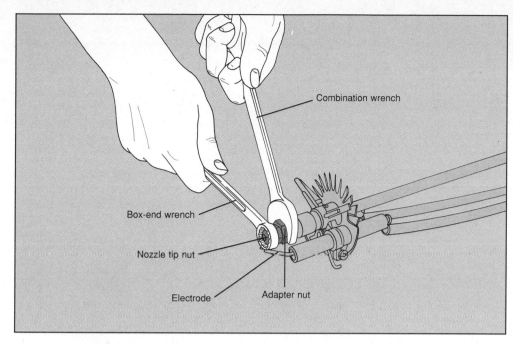

Combination wrench

Box-end wrench

Nozzle tip nut

Electrode

Adapter nut

Removing and replacing the nozzle. Remove the firing assembly *(page 66)*, and locate the two nuts on the nozzle, one near the tip and one at the nozzle oil-line adapter. Taking care not to touch the electrodes or twist the oil line, loosen the nuts with a combination wrench and a box-end wrench, and unscrew the nozzle *(left)*. When choosing a new nozzle, match the specifications stamped on the nozzle tip: firing rate in gallons of oil per hour (GPH); angle of spray in degrees; and spray pattern (identified by a letter of the alphabet). Screw on the new nozzle and reinstall the firing assembly by reversing the sequence on page 66. Have a combustion test done by a professional as soon as possible.

SERVICING THE ELECTRODES

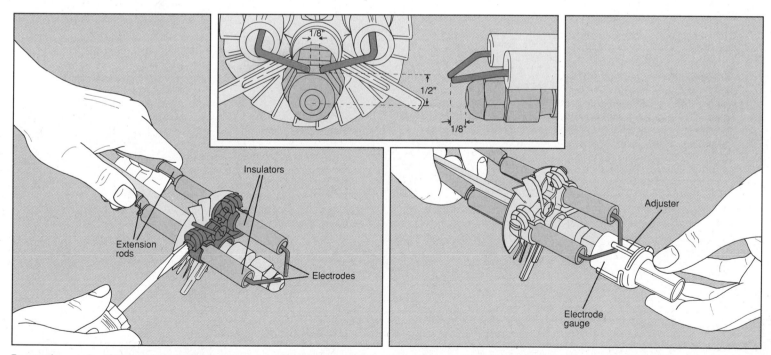

Insulators

Extension rods

Electrodes

Adjuster

Electrode gauge

Inspecting and adjusting the electrodes. Remove the firing assembly *(page 66)*. With power to the burner turned off, examine the electrodes, insulators, and cables or extension rods. Replace cracked insulators or frayed cables. Moisten a cloth with cleaning solvent and wipe soot off the insulators, electrodes and extension rods. Use a dry cloth or brush to wipe out the burner air tube. Vacuum dust from the blower, rotating the blower fan to clean its blades.

Use a ruler to measure between the electrode tips. They should be spaced exactly as specified in the manufacturer's instructions; usually pointed toward each other about 1/8 inch apart, no more

than 1/2 inch above the center of the nozzle tip, and no more than 1/8 inch beyond the front of the nozzle *(inset)*. Adjust the electrode position; if necessary, use an electrode gauge. Check the firing angle, specified in degrees on the nozzle. Match that number to the corresponding number on the electrode gauge; this will tell you which adjuster on the gauge to use. Loosen the screw on the electrode holder *(above, left)*. Gently move the electrodes into place on the electrode gauge *(above, right)*, then tighten the screw. Reinstall the firing assembly and turn on power to the burner.

GAS BURNERS

The gas burner is one of the simplest, most reliable elements in a home heating system. It burns cleaner than an oil burner, is easier to maintain, and promises a constant supply of fuel. Although a gas burner usually fires a forced-air furnace, as shown below, gas-fired boilers with circulating pumps are also in common use.

When the thermostat activates a valve in the combination control, gas flows from a supply line through a manifold, then to burner tubes that mix it with air. The mixture goes to ports where it is ignited by a pilot — or on some newer units, an electric spark igniter. A heat exchanger uses this flame to pro-duce warm air, hot water or steam for circulation through the house. Waste gases go up a vent. The thermocouple stops gas flow if the pilot light goes out or the electric igniter fails.

New technology has improved the efficiency of gas-fired systems. Electric ignition offers two gas-saving alternatives to the conventional pilot: A pilot is ignited electrically, or direct-ignition burners are ignited by a spark from an electrode. High efficiency, or condensing, furnaces operate on a "sealed com-bustion" principle, with no pilot light, burners or conventional vent; an automotive spark plug is used for ignition. Repairs to this type should be done by the gas company.

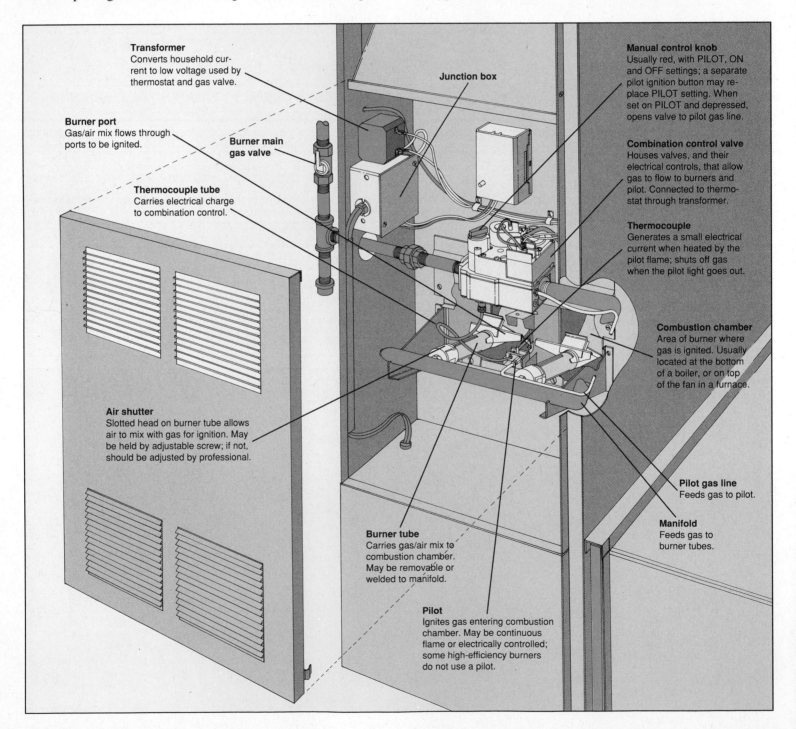

Transformer
Converts household current to low voltage used by thermostat and gas valve.

Burner port
Gas/air mix flows through ports to be ignited.

Burner main gas valve

Thermocouple tube
Carries electrical charge to combination control.

Junction box

Manual control knob
Usually red, with PILOT, ON and OFF settings; a separate pilot ignition button may replace PILOT setting. When set on PILOT and depressed, opens valve to pilot gas line.

Combination control valve
Houses valves, and their electrical controls, that allow gas to flow to burners and pilot. Connected to thermostat through transformer.

Thermocouple
Generates a small electrical current when heated by the pilot flame; shuts off gas when the pilot light goes out.

Combustion chamber
Area of burner where gas is ignited. Usually located at the bottom of a boiler, or on top of the fan in a furnace.

Air shutter
Slotted head on burner tube allows air to mix with gas for ignition. May be held by adjustable screw; if not, should be adjusted by professional.

Burner tube
Carries gas/air mix to combustion chamber. May be removable or welded to manifold.

Pilot gas line
Feeds gas to pilot.

Manifold
Feeds gas to burner tubes.

Pilot
Ignites gas entering combustion chamber. May be continuous flame or electrically controlled; some high-efficiency burners do not use a pilot.

Gas burner maintenance typically involves relighting the pilot *(page 70)*, or servicing the thermocouple *(page 71)* or transformer *(page 73)*. For safety reasons, the combination control and other components that carry gas under pressure should be adjusted or replaced only by the gas company. Gas leaks are rare, but if you smell gas, follow the instructions in the Emergency Guide *(page 9)*.

Your gas burner should be cleaned periodically; how often depends on the burner's age and the type of system it fires. Newer burners require cleaning every few years, but more frequent cleaning is recommended for older models. Prevent rust-causing condensation inside the furnace by turning off the pilot at the end of each heating season.

Before beginning most repairs, turn off power to the burner at the main service panel *(page 134)* or unit disconnect switch *(page 135)*. **Caution:** Some repairs, such as replacing a thermocouple, require shutting off the burner gas valve *(page 72)*. If your gas-fired heating system is not working properly, the cause may be the gas burner itself, or another element of the heating system. Before attempting a repair, also consult the Troubleshooting Guides for System Controls *(page 16)*, and Air Distribution *(page 24)* or Water Distribution *(page 39)*.

TROUBLESHOOTING GUIDE

SYMPTOM	POSSIBLE CAUSE	PROCEDURE
No heat	No power to circuit	Replace fuse or reset circuit breaker *(p. 134)* □○
	Transformer faulty	Test transformer *(p. 73)* ◪●▲ ; replace if necessary
	Pilot light out	Relight pilot *(p. 70)* □○
	Combination control faulty	Call for service
Insufficient heat	Air shutter needs adjustment	Adjust burner air shutter *(p. 72)* □○
	Burner ports clogged	Clean burner ports *(p. 75)* □○
	Gas pressure too low	Call for service
Excessive fuel consumption	Pilot set too high	Adjust pilot *(p. 71)* □○
	Gas pressure too high	Call for service
Pilot does not light or does not stay lit	Pilot orifice dirty or clogged	Clean pilot *(p. 74)* □○
	Thermocouple loose or faulty	Test thermocouple *(p. 71)* □○▲ ; tighten or replace if necessary
	Electric pilot faulty	Call for service
	Combination control faulty	Call for service
Pilot lights, but burner does not ignite	Gas pressure too low	Call for service
	Combination control faulty	Call for service
	Transformer faulty	Test transformer *(p. 73)* ◪●▲ ; replace if necessary
Pilot flame flickers	Pilot set too low or too high	Adjust pilot *(p. 71)* □○
Exploding sound when burner ignites	Pilot set too low	Adjust pilot *(p. 71)* □○
	Pilot orifice dirty or clogged	Clean pilot *(p. 74)* □○
	Gas pressure too low or too high	Call for service
	Pilot light not positioned correctly	Call for service
	Burner ports clogged	Clean burner ports *(p. 75)* □○
Burner takes more than a few seconds to ignite	Burner ports clogged	Clean burner ports *(p. 75)* □○
	Pilot needs adjustment	Adjust pilot *(p. 71)* □○
Burner flame uneven	Burner ports clogged	Clean burner ports *(p. 75)* □○
Burner flame too yellow	Burner dirty	Clean burner *(p. 75)* □○
	Insufficient air for combustion	Provide air from outside by opening vents in furnace room; if problem persists, call for service
	Air shutter opening too small	Adjust air shutter *(p. 72)* □○
	Burners faulty	Call for service
Noisy furnace: rumbling when burners off	Pilot needs adjustment	Adjust pilot *(p. 71)* □○
Noisy furnace: rumbling when burners on	Burner ports clogged	Clean ports *(p. 75)* □○
	Air shutter needs adjustment	Adjust air shutter *(p. 72)* □○

DEGREE OF DIFFICULTY: □ Easy ◪ Moderate ■ Complex
ESTIMATED TIME: ○ Less than 1 hour ◑ 1 to 3 hours ● Over 3 hours ▲ Special tool required

LIGHTING THE PILOT

1 **Removing the front access panel.** Look for screws holding the panel in place, and remove them. Grasp the panel firmly and slide it up, then pull it off *(above)*. Loosen a panel that sticks by rapping it gently at the bottom.

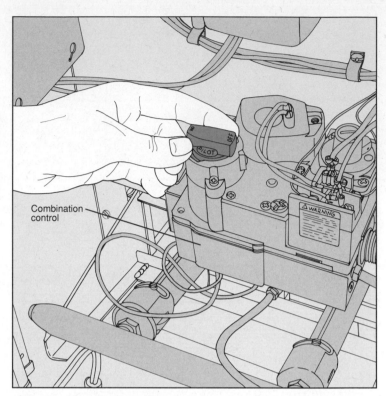

2 **Turning off gas to the pilot.** Set the manual control knob, which is usually red, to the OFF position *(above)* and wait ten minutes for the gas to dissipate before lighting a flame. **Caution:** If the smell of gas persists, do not attempt to relight the pilot; call for service.

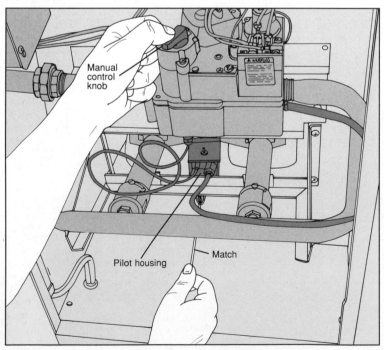

3 **Relighting the pilot.** Follow the manufacturer's instructions for relighting the pilot—usually labeled on, or near, the combination control. If there are no instructions, follow this general procedure: Turn on the gas to the pilot by setting the manual control knob to the PILOT position. While depressing the control knob or pilot ignition button, light the pilot with a long match *(left)* or a lit soda straw. Continue depressing the knob for 30 seconds. Release the control knob; if the pilot goes out, light it again, this time depressing the knob for one minute. If the pilot doesn't stay lit, the thermocouple may be faulty; service the thermocouple *(page 71)*. If the pilot stays lit, check to see that the pilot flame envelops the thermocouple properly; adjust the pilot if necessary *(page 71)*. Turn the manual control knob to the ON position. Replace the front access panel.

CHECKING AND ADJUSTING THE PILOT FLAME

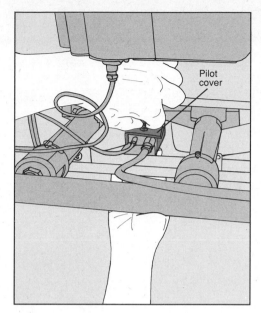

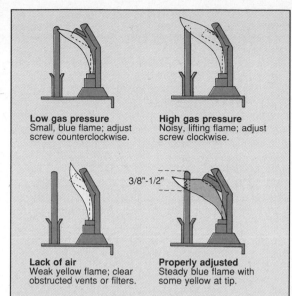

Low gas pressure
Small, blue flame; adjust screw counterclockwise.

High gas pressure
Noisy, lifting flame; adjust screw clockwise.

3/8"-1/2"

Lack of air
Weak yellow flame; clear obstructed vents or filters.

Properly adjusted
Steady blue flame with some yellow at tip.

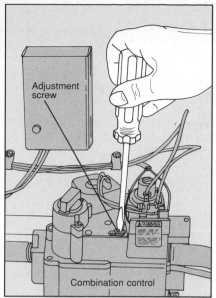

Adjustment screw

Combination control

1 **Accessing the flame.** If you can see the pilot flame clearly, go to step 2. If, however, view of the flame is blocked by a metal cover, first turn off the gas at the manual control knob on the combination control. Wait a few minutes for the cover to cool. Unscrew and remove the cover *(above)*, then turn on the gas and relight the pilot *(page 70)*.

2 **Adjusting the pilot screw.** A properly adjusted flame is blue with some yellow at the tip, and contacts about 3/8 to 1/2 inch of the thermocouple. Compare the flame to the diagrams *(above, left)*, and use the guidelines given to make any necessary adjustments. The adjustment screw on the combination control regulates the amount of gas delivered to the pilot. On some models, the adjustment screw is recessed and covered by a cap screw that must first be removed; on others, the adjustment screw is on the surface of the combination control. Increase the height of the flame by turning the adjustment screw counterclockwise *(above, right)*; decrease it by turning clockwise. Adjust until the flame appears correct. If the flame cannot be correctly adjusted, clean the pilot *(below)*. If you removed a pilot cover in step 1, turn off the gas, replace the cover, then relight the pilot.

ADJUSTING AND REPLACING THE THERMOCOUPLE

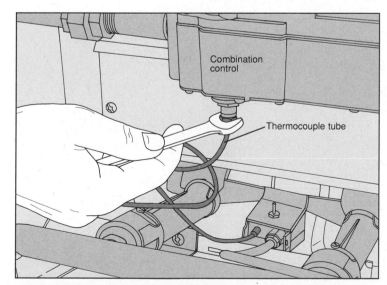

Combination control

Thermocouple tube

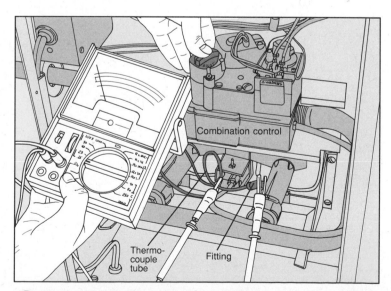

Combination control

Thermocouple tube

Fitting

1 **Removing the thermocouple tube.** The thermocouple generates an electrical charge when it is heated by the pilot flame. To test it, the pilot must remain lit. Turn the manual control knob on the combination control to the PILOT setting and depress the knob or the pilot ignition button. Since it must remain depressed during removal and testing, this will be easier if someone helps you. Detach the thermocouple tube from the combination control by unscrewing the fitting with an open-end wrench *(above)*.

2 **Testing the thermocouple.** Set a multitester to the DCV scale, lowest volt range *(page 137)*. Keeping the manual control knob or button depressed, clip one multitester lead to the end of the thermocouple tube nearest the pilot and the other multitester lead to the fitting on the other end of the tube *(above)*. If the multitester shows a reading, the thermocouple is generating sufficient voltage; put back the thermocouple tube. If there is no reading, release the manual control knob and go to step 3.

ADJUSTING AND REPLACING THE THERMOCOUPLE (continued)

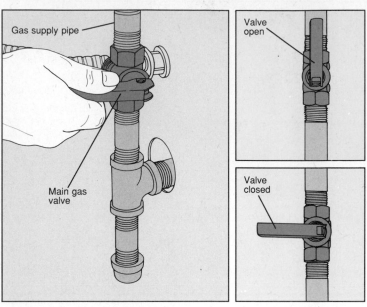

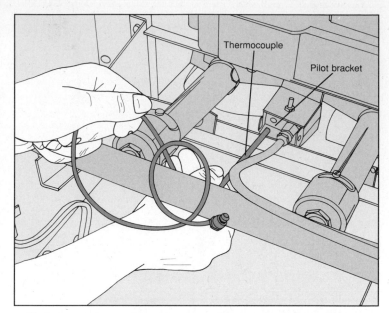

3 **Shutting off the burner main gas valve.** Before replacing the thermocouple, shut off the main gas valve *(above)*, located on the gas supply pipe that leads into the burner. When the handle is parallel to the pipe, the valve is open *(inset, top)*; when the handle is perpendicular to the pipe, the valve is closed *(inset, bottom)*. Turn off power to the burner at the main service panel *(page 134)* or unit disconnect switch *(page 135)*, and allow 30 minutes for metal parts to cool.

4 **Removing and replacing the thermocouple.** Slide the defective thermocouple out of the bracket that holds it in place next to the pilot *(above)*. Use a cloth to clean the fitting on the combination control and screw a new thermocouple tube into the fitting. After tightening it by hand, turn it a quarter turn with an open-end wrench. Insert the thermocouple into the pilot bracket, being careful not to crimp the tubing.

ADJUSTING THE BURNER FLAME

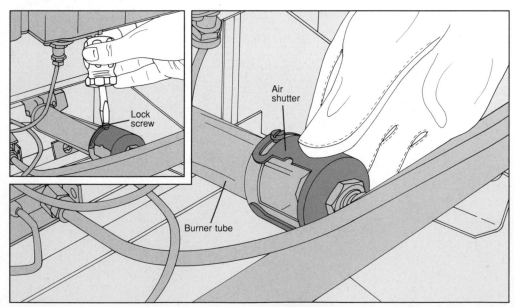

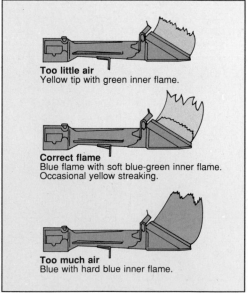

Too little air
Yellow tip with green inner flame.

Correct flame
Blue flame with soft blue-green inner flame. Occasional yellow streaking.

Too much air
Blue with hard blue inner flame.

Adjusting the air shutter. The burner may have an adjustable shutter, held in position by a lock screw on the end of the burner tube; or a fixed air shutter, which must be adjusted by a professional; or no air shutter at all. If your model has an adjustable shutter, turn the thermostat to its highest setting to start the burner and keep it running. Allow five minutes for the burners to ignite and heat up, then remove the burner access panel *(page 70)* and loosen the lock screw *(inset)*. Slowly rotate the shutter open *(above, left)* until the blue base of the flame appears to lift slightly from the burner port surface. Then close the shutter until the flame reseats itself on the surface, and looks correct *(above, right)*. Tighten the lock screw. Repeat this procedure to adjust the remaining burners, then replace the access panel and return the thermostat to its normal setting.

SERVICING THE TRANSFORMER

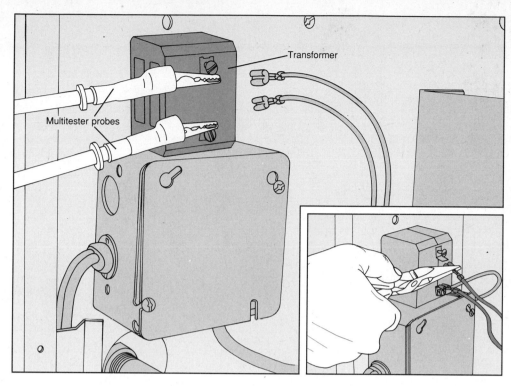

Multitester probes

Transformer

1 **Testing the transformer.** Turn off power to the burner at the main service panel *(page 134)* or unit disconnect switch *(page 135)*. Remove the access panel *(page 70)* and locate the two wires that connect the transformer to the combination control. Using long-nose pliers, gently wiggle these two wires off their terminals on the transformer *(inset)*. Also detach the ground wire. Set a multitester to the ACV scale, 50-volt range. Clip one multitester probe to each transformer terminal *(left)*. Without touching the probes or transformer, turn on power to the burner. If the multitester reads about 12 volts or 24 volts, the transformer is sending sufficient voltage; reattach the wires. If there is no reading, turn off power to the burner and go to step 2.

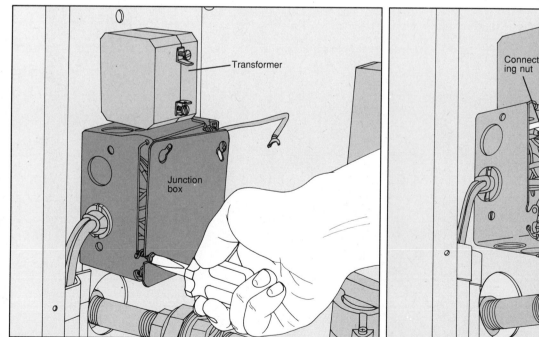

Transformer

Junction box

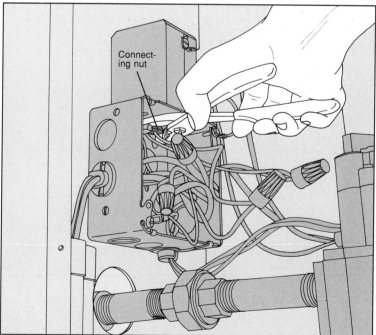

Connecting nut

2 **Opening the junction box.** With power to the burner turned off, unscrew *(above, left)* and remove the junction box cover. Inside the junction box, locate the two wires leading to the transformer; inspect the wire cap connections that join them to other wires *(page 139)*. Reconnect the wires if they are loose *(page 139)* and test again *(step 1)*. If the connections do not seem to be the problem, use pliers to unscrew the nut connecting the transformer to the junction box *(above, right)* and replace the transformer *(step 3)*.

SERVICING THE TRANSFORMER (continued)

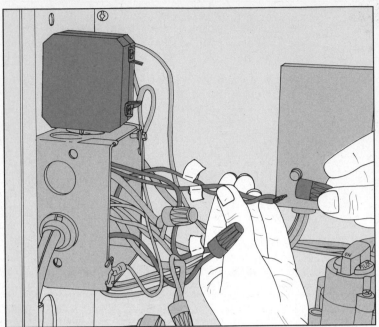
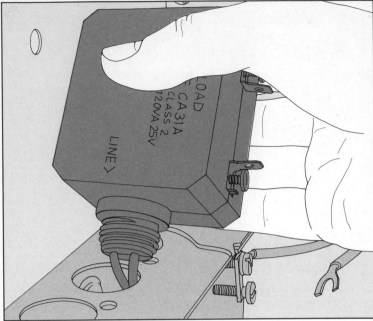

3 **Removing and replacing the transformer.** Remove the two wire caps securing the transformer wires *(above, left)*, first labeling all the wires as "bundle A" or "bundle B" for correct reconnection *(page 135)*. Unwind and free the wires. Lift the transformer off the junction box *(above, right)*. Install an identical transformer. Tighten the nut and re-connect the transformer wires with their labeled counterparts. Screw on the junction box cover and turn on power to the burner.

CLEANING THE PILOT

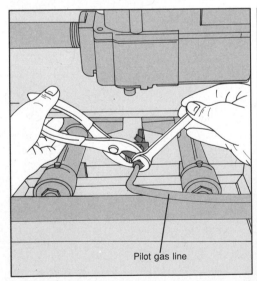
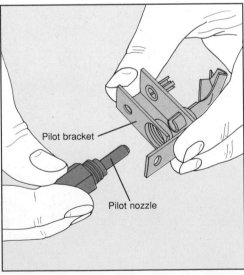
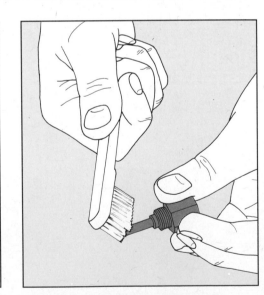

Pilot bracket

Pilot nozzle

Pilot gas line

1 **Disassembling the pilot.** Shut off the gas to the burner by turning the main gas valve to the OFF position *(page 72)*. Turn off power to the burner at the main service panel *(page 134)* or unit disconnect switch *(page 135)*. If the burner has been running, wait 30 minutes to allow metal parts to cool. Disconnect and remove the thermocouple tube *(page 71)*. With a pair of pliers, steady the gas line connecting the pilot to the combination control; be careful not to bend or damage the line. With an open-end wrench in your other hand, loosen the nut attaching the pilot gas line *(above, left)*. Unscrew and remove the bracket holding the pilot/thermocouple assembly in place. Carefully unscrew the pilot nozzle from the bracket and pull the two apart *(above, right)*.

2 **Cleaning the pilot nozzle.** Brush surface dirt off the pilot nozzle with a toothbrush *(above)*. Use a soft wire to dislodge deposits from inside the pilot, being careful not to damage or chip it. Reassemble the pilot assembly and reinstall it in the burner. Turn on the gas and power to the burner and relight the pilot *(page 70)*.

CLEANING THE BURNERS AND SPUDS

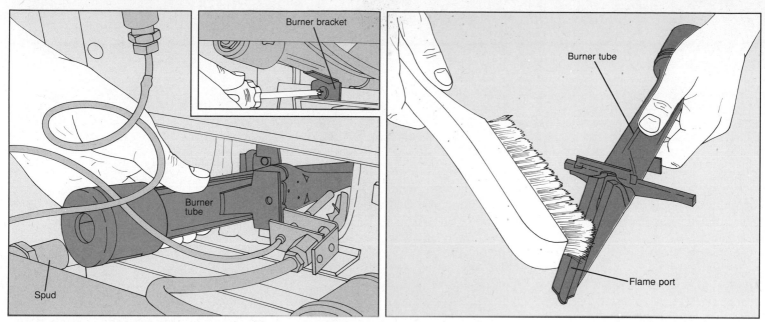

1 **Removing and cleaning the burner tubes.** Shut off the main gas valve *(page 72)* and turn off power to the burner at the main service panel *(page 134)* or unit disconnect switch *(page 135)*. If the burner tubes are housed in a metal drawer, pull out the drawer, then clean the ports gently with a brush. If the burner tubes are the removable type, such as the model shown, there will usually be a screw attaching each burner tube to the retaining bracket just underneath the tubes; remove these screws *(inset)*. On some models, the entire pilot assembly must be removed to gain access to the burner tubes. When the burner tube is free of the retaining bracket, gently twist it forward off the spud, then pull it out of the burner *(above, left)*. Rust and soot can accumulate on the exterior of the burner tubes; clean the tubes thoroughly with a stiff-bristled brush, paying special attention to the ports *(above, right)*. **Caution:** Clean gently to avoid enlarging or damaging the burner ports.

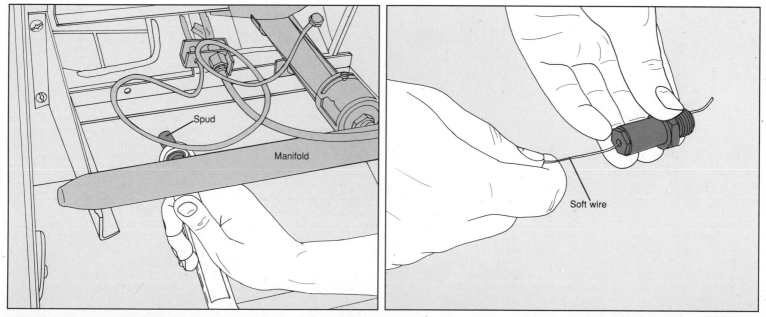

2 **Removing and cleaning the spuds.** Use a box-end wrench to unscrew the spud from the manifold *(above, left)*; remove the spud. Run a soft wire through the spud opening to clear it, being careful not to chip or enlarge the opening *(above, right)*. Gently tighten the spud onto the manifold, then reinstall the burner assembly, reversing the procedure here.

ELECTRIC FURNACES

The heat produced by an electric furnace is cleaner than in oil- or gas-burning types, because no combustion takes place. In areas where electricity is still affordable, an electric furnace will heat efficiently, provided the house is well insulated. The electric furnace may be linked to a heat pump *(page 90)* or central air conditioning *(page 102)*, and is almost always used with a forced-air distribution system. The unit draws cool air from the return registers and blows it over three to six electric heating elements. A blower in the furnace housing *(Air Distribution, page 24)* sends the heated air through the plenum and into the rooms of the house via the ducts.

The home's 240-volt electrical current enters the unit from the main service panel and unit disconnect switch, then flows to the transformer and blower. The transformer steps down the line voltage to 24 volts, which flows to the control terminal block — the furnace's link to the thermostat. When the thermostat calls for heat, the control terminal block directs low voltage to the heat relay. The heat relay signals the first sequencer relay's contacts to close, allowing line voltage to flow to the first element. This sequencer then signals the next sequencer and the process continues, turning on as many elements as are needed to satisfy the thermostat. The blower also turns on, and sends heated air through the system. On some larger units, electrical current in the element circuit also flows through cartridge fuses to protect against overload.

Apart from filter, blower and humidifier maintenance *(Air Distribution, page 24)*, or special maintenance required by central air conditioning, the electric furnace is virtually maintenance free. Once a year, access the control box and check the wires and terminal connections; replace burned or damaged wires, and inspect the replacements periodically. If the damage recurs, call an electrician.

Troubleshooting an electric furnace may be as simple as testing and replacing a fuse *(page 80)* or as involved as replacing a heating element, which may require disconnecting a number of wires that are in the way. When disconnecting more than one wire, make sure to label the wires first, to ensure correct reconnection. The components in your electric furnace may vary from those pictured at right. If in doubt about how to proceed with a repair, call for professional service.

Follow all safety instructions in this chapter and in the Emergency Guide *(page 8)*. Before beginning any repair or inspection, turn off power to the heating and cooling system at the main service panel *(page 79)*. Heating problems may also be caused by a component in the air distribution system, or by the thermostat. Consult the Troubleshooting Guides in System Controls *(page 16)* and Air Distribution *(page 24)*.

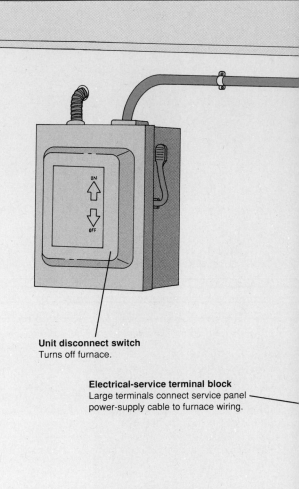

Unit disconnect switch
Turns off furnace.

Electrical-service terminal block
Large terminals connect service panel power-supply cable to furnace wiring.

Control terminal block
Connects thermostat to unit's electrical components.

Sequencer relays. With a heating filament and a bi-metal switch, the sequencer relay uses one electrical circuit to turn on another circuit. When the thermostat calls for heat, 24-volt electrical current flows through the filament of the first relay. After several seconds, the heat created by the current flow causes the bimetal switch in the relay to expand, switching on 240-volt current to its heating element. At the same time, the relay allows 24-volt current to pass to the next sequencer relay in the series, which turns on the 240-volt current to its heating element. This process continues, turning on elements one at a time until the demands of the thermostat have been satisfied. To protect the household electrical supply against overload, each sequencer relay has a delayed reaction, switching on its element several seconds after the previous one. The wiring of sequencer relays varies according to the model they serve. To avoid any risk of disconnecting the wrong wires, call for service if your troubleshooting points to a faulty sequencer relay.

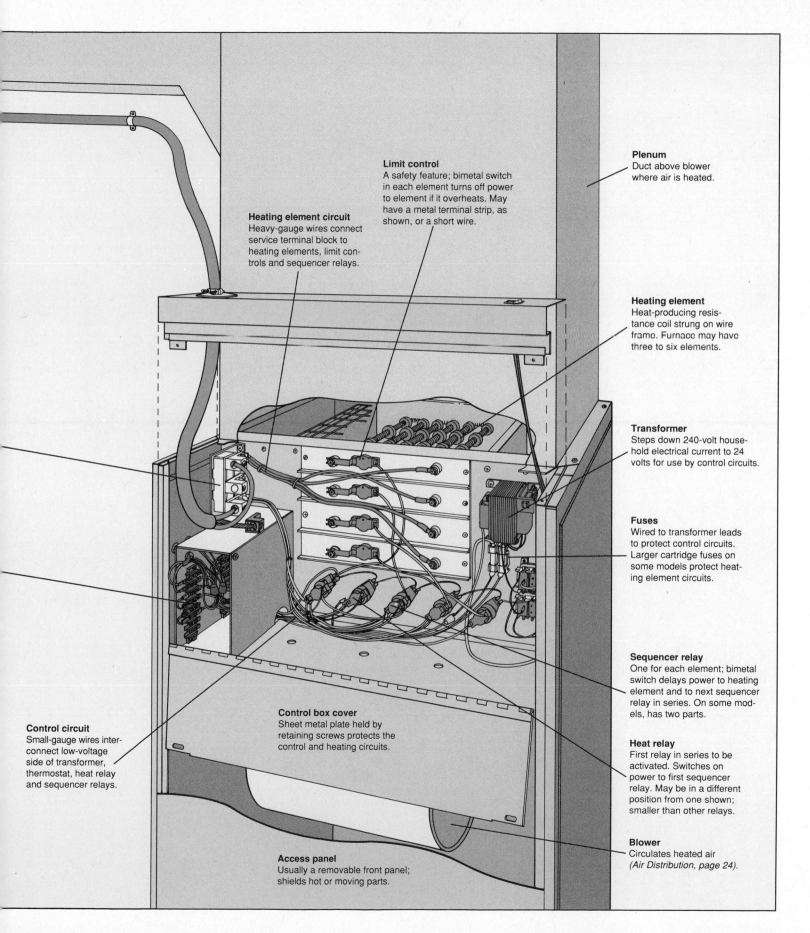

Limit control
A safety feature; bimetal switch in each element turns off power to element if it overheats. May have a metal terminal strip, as shown, or a short wire.

Heating element circuit
Heavy-gauge wires connect service terminal block to heating elements, limit controls and sequencer relays.

Plenum
Duct above blower where air is heated.

Heating element
Heat-producing resistance coil strung on wire frame. Furnace may have three to six elements.

Transformer
Steps down 240-volt household electrical current to 24 volts for use by control circuits.

Fuses
Wired to transformer leads to protect control circuits. Larger cartridge fuses on some models protect heating element circuits.

Sequencer relay
One for each element; bimetal switch delays power to heating element and to next sequencer relay in series. On some models, has two parts.

Heat relay
First relay in series to be activated. Switches on power to first sequencer relay. May be in a different position from one shown; smaller than other relays.

Control box cover
Sheet metal plate held by retaining screws protects the control and heating circuits.

Control circuit
Small-gauge wires interconnect low-voltage side of transformer, thermostat, heat relay and sequencer relays.

Blower
Circulates heated air (Air Distribution, page 24).

Access panel
Usually a removable front panel; shields hot or moving parts.

TROUBLESHOOTING GUIDE

SYMPTOM	POSSIBLE CAUSE	PROCEDURE
No heat	Circuit breaker tripped or fuse blown	Reset breaker or replace fuse *(p. 134)* ☐○
	Wires broken or terminal connectors loose or faulty	Inspect furnace wiring. Replace wires or install new connectors *(p. 138)* ▰◑; tighten wire connections *(p. 80)* ☐○
	Wires burned (may be accompanied by burned paint on furnace housing)	Inspect furnace wiring. Replace wires *(p. 138)* ▰◑; if problem recurs, have electrician check for aluminum house wiring and replace it
	Control or heating circuit fuse faulty	Test fuse and replace *(p. 80)* ☐○▲ if necessary
	Transformer faulty	Test transformer and replace *(p. 81)* ▰◑▲ if necessary
	Limit controls faulty	Identify faulty element circuit *(p. 82)* ▰○▲; test limit control and replace *(p. 82)* ▰◑▲ if necessary
	Heating elements burned out, due to dirty air filter	Replace air filter *(Air Distribution, p. 26)* ☐○. Identify faulty element circuits *(p. 82)* ▰○▲; test heating elements and replace *(p. 83)* ▰◑▲ if necessary
	Heating elements faulty	Identify faulty element circuit *(p. 82)* ▰○▲; test heating elements and replace *(p. 83)* ▰◑▲ if necessary
	Heat relay faulty	Call for service
	Sequencer relay faulty	Call for service
Intermittent or insufficient heat	Wire connections loose	Inspect furnace wiring. Replace loose spade lugs *(p.139)* ☐○; tighten wire connections *(p. 80)* ☐○
	Wires burned (may be accompanied by burned paint on furnace housing)	Inspect furnace wiring. Replace wires *(p. 138)* ▰◑; if problem recurs, have electrician check for aluminum house wiring and replace it
	Thermostat anticipator not adjusted for electric furnace	Adjust thermostat anticipator *(System Controls, p. 19)* ☐○
	Old-style thermostat does not have anticipator	Replace thermostat with new style that has anticipator *(System Controls, p. 16)* ☐○
	Limit control faulty	Identify faulty element circuit *(p. 82)* ▰○▲; test limit control and replace *(p. 82)* ▰◑▲ if necessary
	Heating element faulty	Identify faulty element circuit *(p. 82)* ▰○▲; test heating element and replace *(p. 83)* ▰◑▲ if necessary
	Sequencer relay faulty	Call for service

DEGREE OF DIFFICULTY: ☐ Easy ▰ Moderate ■ Complex
ESTIMATED TIME: ○ Less than 1 hour ◑ 1 to 3 hours ● Over 3 hours   ▲ Special tool required

---

TROUBLESHOOTING FOR ELECTRICAL PROBLEMS

Inspect your furnace periodically for signs of overheating electrical circuits; burned wire insulation and scorched paint on the housing are common clues. You can usually solve the problem by tightening wire connections or replacing damaged wires *(page 138)*.

If your house wiring is aluminum, however, the problem is more fundamental. Many homes built or enlarged between 1965 and 1972 have branch circuits of aluminum wiring. Over a 10-year period, some 500 fires were attributed to this type of wiring. Two factors make aluminum wiring potentially hazardous: Corrosion of the aluminum wire ends can cause high resistance to current at electrical terminals, generating a great deal of heat. In addition, since aluminum expands and contracts more than other metals when heated and cooled, aluminum wiring tends to wiggle loose from its terminals, adding to resistance and overheating. Especially in a vibrating furnace, which cycles on and off continually, it is almost impossible to maintain a good connection with aluminum wiring.

Check for aluminum wiring entering the furnace at the electrical-service terminal block; look for loose or corroded terminals, and burned insulation or paint. If the electrical system in your home uses aluminum wiring, consider having an electrician replace it.

Another tell-tale troubleshooting symptom is short cycling of the furnace. A furnace that turns on and off repeatedly may be linked to a thermostat that is not calibrated properly or that is simply the wrong thermostat for the system. Many of the electric furnaces in use today were installed as replacements for oil or gas furnaces, and the original thermostats were not adjusted or replaced accordingly. For efficient operation, an electric furnace requires a thermostat with a heat anticipator; if your thermostat does not have one, replace it with a model that does *(page 22)*. If your thermostat does have a heat anticipator, it may not have been readjusted when the electric furnace was installed. Check the amperage rating of the furnace, and adjust the thermostat anticipator if necessary *(page 19)*.

TURNING OFF POWER TO THE ELECTRIC FURNACE

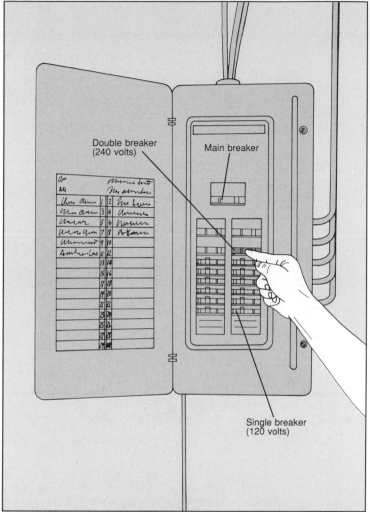

Double breaker
(240 volts)

Main breaker

Single breaker
(120 volts)

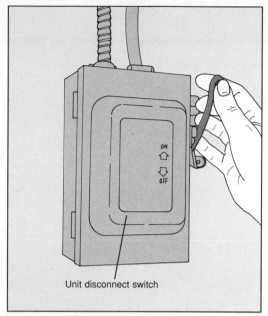

Unit disconnect switch

Cutting power at the main service panel and unit disconnect switch. Before servicing an electric furnace, turn off power to the furnace. Look on the service panel—usually below the main breaker or fuse blocks *(page 134)*—and identify the breaker or fuses that protect your unit *(left)*. If this information is not written on the service panel, turn off each breaker or remove each fuse until power to the furnace is off; then take the opportunity to label the service panel for future reference. Once you have identified the proper breaker, move it to the OFF position; it will stay in that position or spring back to an intermediate position. If your system has fuses, there may be two that control the furnace. Twist them out counterclockwise. If the furnace also has a separate unit disconnect switch located on the wall near it, turn it off with one hand as an extra safety precaution *(above)*.

ACCESSING THE INTERNAL COMPONENTS

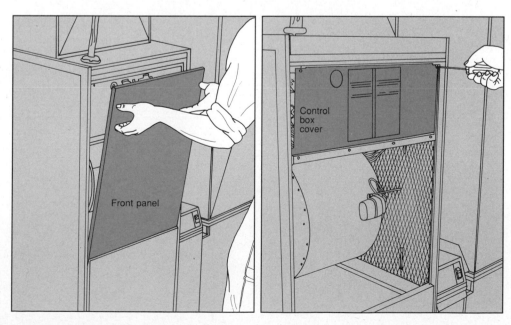

Front panel

Control
box
cover

Removing the access panels. Turn off power to the furnace at the main service panel and unit disconnect switch *(step above)*. Grasp the slotted handles on the front panel and pull the panel away with a sharp upward tug *(far left)*, exposing the control panel. Place the panel out of the way. Next, unscrew the retaining screws from the control box cover to gain access to the internal wiring and components.

SERVICING WIRE CONNECTIONS

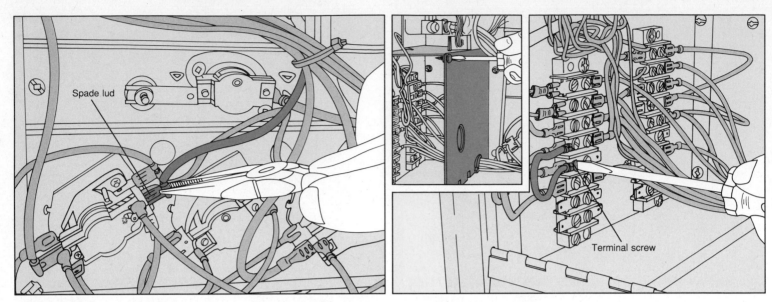

Inspecting wire connections. Turn off power to the furnace at the main service panel and unit disconnect switch, and access the control box *(page 79)*. Visually inspect the wiring in the control box for loose connectors, frayed wires, cracked insulation, and signs of overheating or burning. Replace damaged wires *(page 138)*. Use long-nose pliers to tug gently on wires connecting individual components inside the control box *(above, left)*; tighten loose connections. Replace loose-fitting or damaged spade lugs *(page 139)*. Remove the control terminal block cover *(inset)*, and use a screwdriver to tighten terminal screws that may have worked loose due to blower vibration *(above, right)*.

TESTING AND REPLACING FUSES

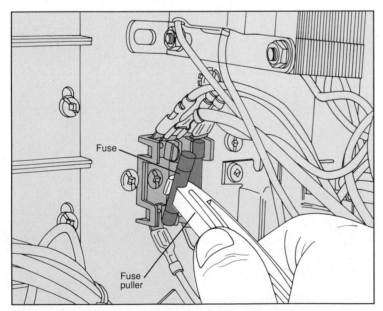

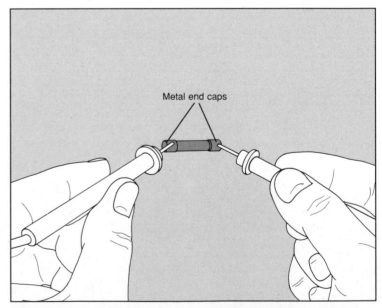

1 **Removing control circuit and heating circuit fuses.** Turn off power to the furnace at the main service panel and unit disconnect switch, and access the control box *(page 79)*. Locate the fuses, in a block wired to the transformer leads, such as the one shown, or larger cartridge fuses connected to a panel in the control box. (Some units have both types.) Use a fuse puller to hook the center of the fuse cartridge and, with a quick tug, free the fuse from its block *(above)*.

2 **Testing a fuse.** Set a multitester to the RX1 setting to test continuity *(page 136)*. Touch one multitester probe to each end cap of the fuse *(above)*; test a larger cartridge fuse the same way. If the multitester shows continuity, the fuse is good; test all other fuses in the unit. If a fuse has no continuity, purchase an exact replacement—available at a hardware store—and install it in its block, or on its panel. Restore the power. If a replacement fuse blows, turn off power to the furnace and call for service.

SERVICING THE TRANSFORMER

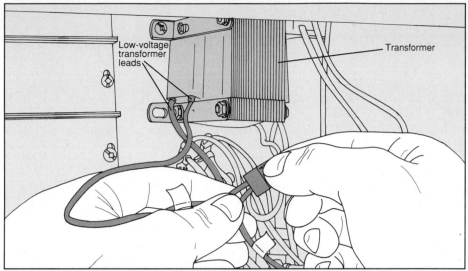
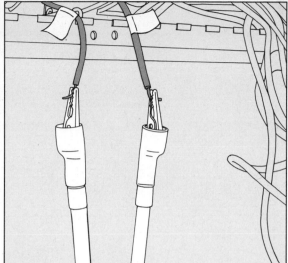

1 **Testing the transformer output.** Turn off power to the furnace at the main service panel and unit disconnect switch, and access the control box *(page 79)*. Locate the low-voltage wire leads running to the control terminal block from the transformer, and label them for correct reconnection *(page 135)*. Disconnect the low-voltage leads by unscrewing their wire caps *(above, left)*. Set a multitester to the ACV setting and turn the dial to the 50-volt range to test voltage *(page 137)*. Using alligator clips, attach a multitester probe to each disconnected lead *(above, right)*. Being careful not to touch the furnace or any wire, turn on power to the furnace, note the multitester reading, then turn off the power. The tester should read about 24 volts. If not, test the transformer windings *(step 2)*.

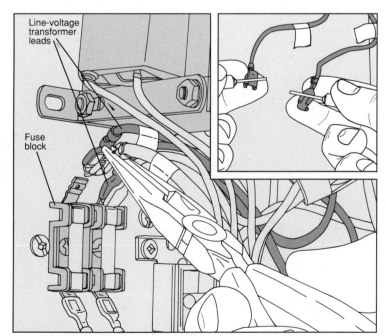
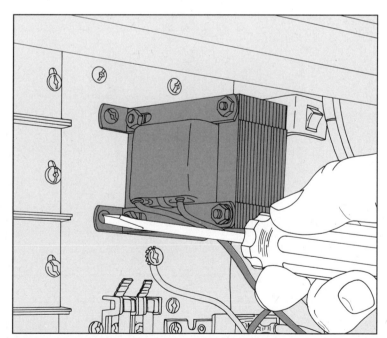

2 **Testing the transformer windings.** With power to the furnace turned off, trace the transformer's line-voltage wire leads to the control fuse block; label the wires for correct reconnection *(page 135)*. Use long-nose pliers to detach their spade lugs from the terminals *(above)*. Set a multitester to the RX1K setting to test continuity *(page 136)*. Touch a multitester probe to each lead *(inset)*. If the multitester shows continuity, test the other transformer leads *(step 1)* the same way. If either test shows no continuity, replace the transformer *(next step)*.

3 **Replacing the transformer.** With power to the furnace turned off, use a screwdriver to remove the retaining screws securing the transformer in the control box; support the transformer with one hand and lift it out of the furnace. Purchase an identical replacement transformer from an electrical supplies dealer or from the manufacturer, and screw it in place. Connect the wires to their terminals, and restore the power.

IDENTIFYING A FAULTY HEATING ELEMENT CIRCUIT

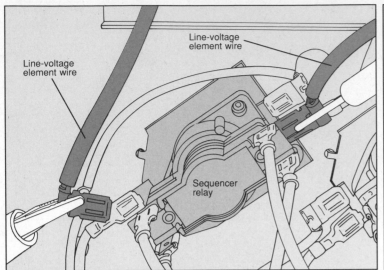

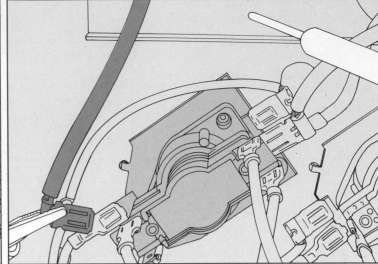

Testing the element circuits. Turn off power to the furnace at the main service panel and unit disconnect switch, and access the internal components *(page 79)*. Conduct two tests for each element. First, trace the line-voltage wire that leads from the element to the sequencer relay; then find the similar line-voltage wire connected to the other side of the relay. Use long-nose pliers to disconnect one of the wires. Set a multitester to the RX1K setting to test continuity *(page 136)*. Attach one alligator clip to the disconnected wire end and touch one probe to the other wire's terminal on the sequencer *(above, left)*. The multitester should show continuity. Next, test for ground by touching one probe or clip to the detached wire and the other probe to bare metal inside the control box *(above, right)*; there should be no continuity. If the element fails either test, test the limit control for that circuit *(below)*. Otherwise, reconnect the wires and test the other element circuits. If all element circuits test OK, a sequencer relay may be faulty; call for service.

TESTING AND REPLACING THE LIMIT CONTROLS

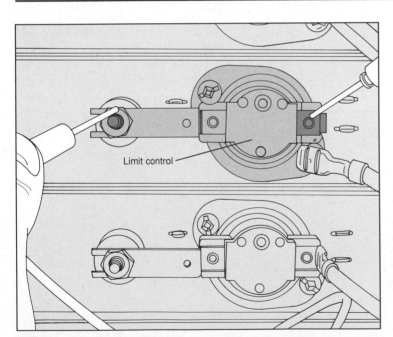

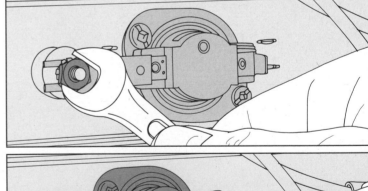

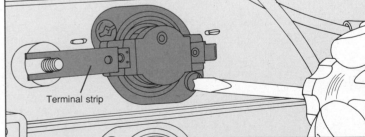

1 **Testing a limit control.** With power to the furnace turned off *(page 79)*, test the limit control on the faulty element circuit *(step above)*. Use long-nose pliers to detach a wire from the limit control. Set a multitester to the RX1K setting to test continuity *(page 136)*. Touch a multitester probe to the terminal on each side of the limit control *(above)*; the model shown above shares a terminal with the heating element. If the multitester does not show continuity for the limit control, replace it *(next step)*. If the limit control tests OK, test the element *(page 83)*.

2 **Replacing the limit control.** Label the limit control wires for correct reconnection *(page 135)*. Using an open-end wrench, unscrew the nut securing the terminal strip to the element terminal *(above, top)*. On models with a short wire instead, pull off the connector. Remove the retaining screws holding the limit control bracket *(above, bottom)*, and pull out the limit control. Purchase an identical replacement from a heating and cooling supplies dealer. Install the new part, reversing these instructions. Reassemble the unit and restore the power.

SERVICING THE HEATING ELEMENT

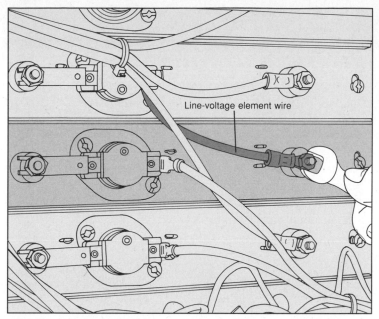

Line-voltage element wire

1 **Disconnecting the heating element from its circuit.** With power to the furnace turned off *(page 79)*, detach one of the wires from the heating element in the faulty element circuit *(page 82)*: Using an open-end wrench, unscrew the nut securing the line-voltage element wire to the heating element terminal *(above)* and remove the wire.

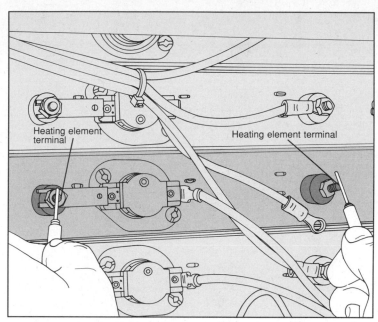

Heating element terminal

Heating element terminal

2 **Testing the element.** With power to the unit off, set the multi-tester to the RX1K setting to test continuity *(page 136)*. Touch a probe to each element terminal *(above)*. The multitester should show continuity; if not, replace the element *(step 3)*. If the element has continuity, and all other components in its circuit test OK *(page 82)*, a sequencer relay may be faulty; call for service.

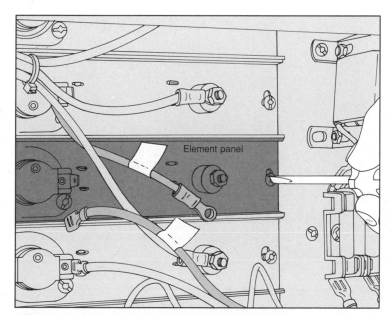

Element panel

3 **Removing the element.** With power to the unit off, label all wires connected to the faulty heating element *(page 135)*. Detach any wire still connected to the element's terminals, then use a screwdriver to remove the retaining screws holding the element panel in place *(above)*.

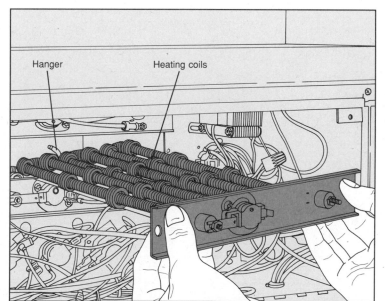

Hanger   Heating coils

4 **Replacing the element.** Grasp the element panel and pull it out *(above)*, working it back and forth gently to release it; avoid striking other elements. The hanger at the back of the element will drop out of its hole in the plenum. Purchase an identical replacement element from a heating and cooling supplies dealer or the manufacturer, and slide it into its slot. Insert the hanger into its hole in the plenum. Press the element panel in place, align it with the screw holes and replace the screws. Reattach all wires to the element, reassemble the unit and restore the power.

BASEBOARD HEATERS

A baseboard heater is a long, narrow unit mounted at floor level. The heating element inside uses electrical current to warm air that moves over it; fins surrounding the element release the heat into the room. Because the heated air rises and then falls as it cools, no fans are required to keep the warm air circulating—provided the air flow is not obstructed by furniture, curtains or dust, and the house is well insulated.

Each baseboard heater is wired directly to the household electrical supply, usually 240 volts, and is controlled by either an internal thermostat, as shown below, or by a wall thermostat *(page 16)*. A limit control switches off the unit if it overheats. Often, this happens simply because the heating element is dirty. Although most limit controls reset themselves automatically, some models have a manual reset button; reset it after cleaning the element *(page 86)*.

If a baseboard heater is faulty, first check the thermostat and limit control, which are easier and less expensive to replace than the heating element. If the heating element proves faulty, you may wish to replace the entire unit; a replacement element may cost nearly as much as a new heater.

Follow all safety instructions in this chapter and in the Emergency Guide *(page 8)*. Before beginning work, turn off power to the baseboard heater circuit at the main service panel *(page 134)*, and test to make sure it is off *(page 85)*. For baseboard heaters that use wall thermostats, also consult the Troubleshooting Guide for System Controls *(page 16)*. Keep baseboard heaters free from obstructions; flammable materials that rest against, or fall into, a unit pose a fire hazard. If you detect a burning smell, turn off power, remove the element panel *(page 86)* and check for objects inside the heater.

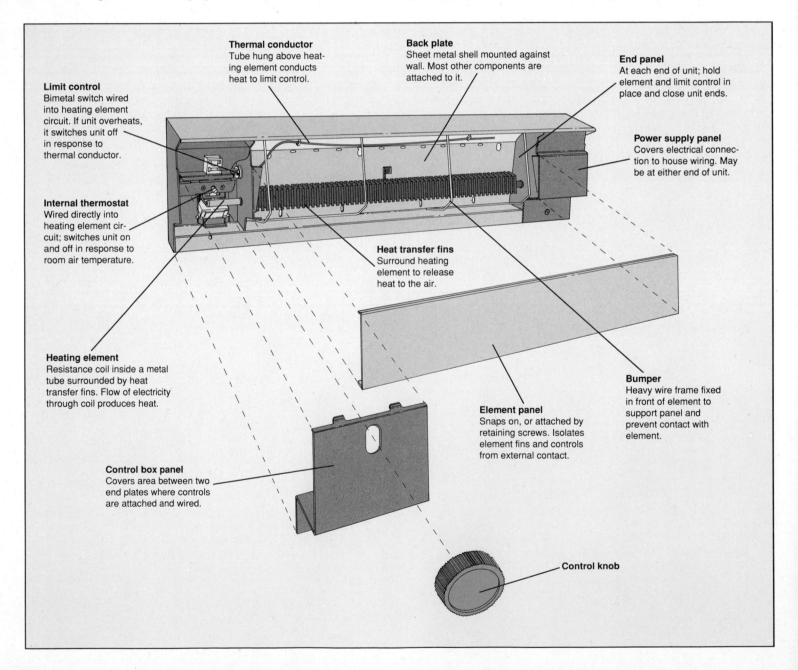

Thermal conductor
Tube hung above heating element conducts heat to limit control.

Back plate
Sheet metal shell mounted against wall. Most other components are attached to it.

End panel
At each end of unit; hold element and limit control in place and close unit ends.

Limit control
Bimetal switch wired into heating element circuit. If unit overheats, it switches unit off in response to thermal conductor.

Power supply panel
Covers electrical connection to house wiring. May be at either end of unit.

Internal thermostat
Wired directly into heating element circuit; switches unit on and off in response to room air temperature.

Heat transfer fins
Surround heating element to release heat to the air.

Heating element
Resistance coil inside a metal tube surrounded by heat transfer fins. Flow of electricity through coil produces heat.

Element panel
Snaps on, or attached by retaining screws. Isolates element fins and controls from external contact.

Bumper
Heavy wire frame fixed in front of element to support panel and prevent contact with element.

Control box panel
Covers area between two end plates where controls are attached and wired.

Control knob

TROUBLESHOOTING GUIDE

SYMPTOM	POSSIBLE CAUSE	PROCEDURE
No heat	No power to system	Replace fuse or reset circuit breaker *(p. 134)* □○
	Internal thermostat faulty	Test internal thermostat and replace *(p. 86)* ▭◖▲ if necessary
	Air supply blocked	Move furniture, curtains or other obstructions
	Dirt or foreign object in heat transfer fins causing limit control to turn off unit	Clean heating element *(p. 86)* □○
	Limit control faulty	Test limit control and replace *(p. 87)* ▭◖▲ if necessary
	Heating element faulty	Test heating element and replace element *(p. 88)* ▭◖▲ or replace entire unit *(p. 89)* ▭◖ if necessary
Intermittent or insufficient heat	Air supply blocked	Move furniture, curtains or other obstructions
	Dirt or foreign object in heat transfer fins causing limit control to turn off unit	Clean heating element *(p. 86)* □○
	Limit control faulty	Test limit control and replace *(p. 87)* ▭◖▲ if necessary
	Internal thermostat faulty	Test internal thermostat and replace *(p. 86)* ▭◖▲ if necessary
	Heating element faulty	Test heating element and replace element *(p. 88)* ▭◖▲ or replace entire unit *(p. 89)* ▭◖ if necessary

DEGREE OF DIFFICULTY: □ **Easy** ▭ **Moderate** ■ **Complex**
ESTIMATED TIME: ○ **Less than 1 hour** ◖ **1 to 3 hours** ● **Over 3 hours**   ▲ **Special tool required**

CHECKING THAT POWER IS TURNED OFF

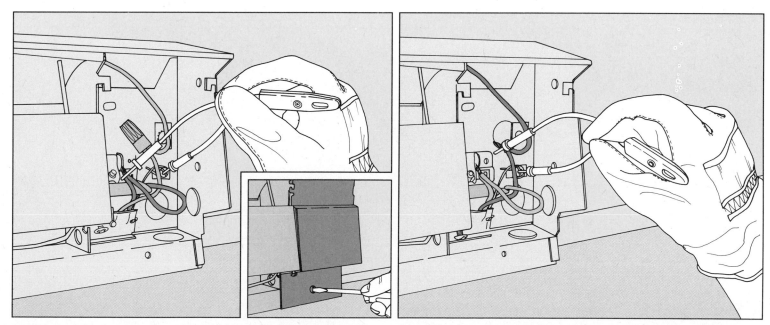

Testing to make sure power is off. At the beginning of each repair, turn off power to the baseboard heater at the main service panel *(page 134)*. Use a screwdriver to remove the power supply panel *(inset)*. Conduct three tests, wearing insulated gloves and taking care not to touch any bare wires. First, unscrew the wire cap connecting the black line-voltage wire to one of the baseboard element wires. Working one-handed, hold a voltage tester by the insulation and touch one tester probe to the uncapped wires and the other tester probe to the metal chassis *(above, left)*. Next, remove the other wire cap and repeat the test with its wires and the chassis *(above, right)*. Finally, touch a tester probe to each uncapped wire connection If the power is indeed off, the tester will not glow in any test. If it does, return to the main service panel and flip off the correct circuit breaker; repeat the voltage test to make sure the power is off.

CLEANING THE HEATING ELEMENT

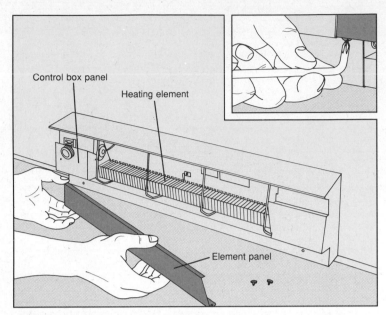

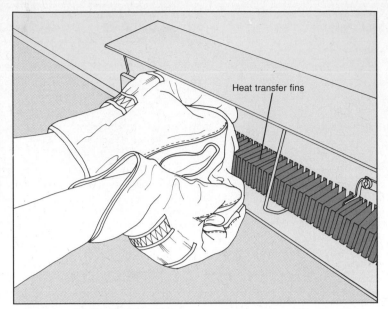

1 **Removing the element panel.** Turn off power to the baseboard heater circuit at the main service panel *(page 134)*, then test to make sure it is off *(page 85)*. If your unit has a separate control box panel and power supply panel at the ends, leave them in place. Remove retaining screws holding the element panel in place, using an angled screwdriver if necessary *(inset)*. Lift off the element panel carefully to avoid bending it *(above)*.

2 **Cleaning the element.** With power turned off, remove accumulated dust by brushing the heat transfer fins with a dry paintbrush. If dust buildup is heavy, wear gloves for protection from sharp edges and wipe the surface of each fin with a damp rag *(above)*. Vacuum away dislodged dirt. If fins are bent, straighten them with small pliers. If the limit control has a manual reset button, remove the control box panel *(below)* and push the reset button. Reassemble the unit and restore power.

SERVICING AN INTERNAL THERMOSTAT

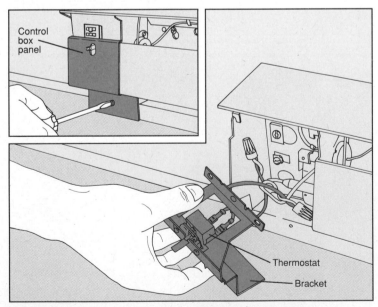

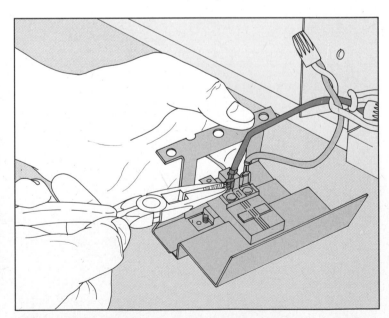

1 **Removing the control box panel and thermostat.** Turn off power to the baseboard heater circuit at the main service panel *(page 134)*, then test to make sure it is off *(page 85)*. Remove the retaining screws holding the control box panel *(inset)* and lift it off. In most cases, the thermostat is mounted on a bracket; unscrew the bracket retaining screws and pull the thermostat and bracket out of the control box *(above)*.

2 **Removing the wire leads.** Disconnect one thermostat lead from its terminal by grasping the spade connector with long-nose pliers, and gently wiggling it off the terminal *(above)*. Some thermostats have screw terminals; unscrew the terminal to disconnect the lead.

SERVICING AN INTERNAL THERMOSTAT (continued)

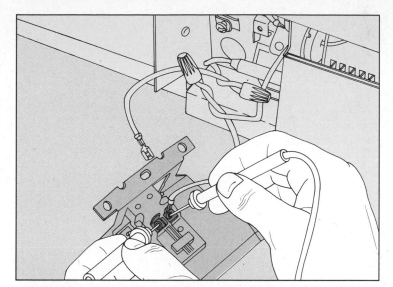

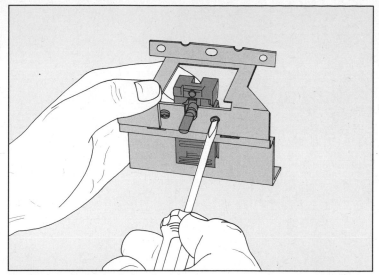

3 **Testing the thermostat.** Set the control knob to LOW or OFF, then turn it gradually toward the highest setting. When the contacts close—usually several degrees above room temperature—you should hear a click. If not, set a multitester to the RX1K setting to test for continuity *(page 136)*. With power to the baseboard heater circuit off, touch a multitester probe to each thermostat terminal *(above)*. Turn the dial from LOW to HIGH. The tester should show no continuity at first, then should jump to continuity at the room temperature setting or above. If not, replace it.

4 **Replacing the thermostat.** Before buying an identical replacement thermostat, check whether it clicks *(step 3)* in the store. With power to the baseboard heater circuit off, unscrew the retaining screws *(above)*; replace the old thermostat with the new one, transferring the wires one at a time to ensure correct connection. Reassemble the heater and restore power.

TESTING AND REPLACING THE LIMIT CONTROL

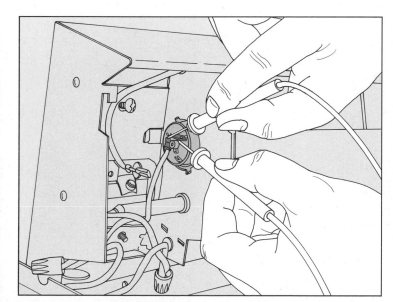

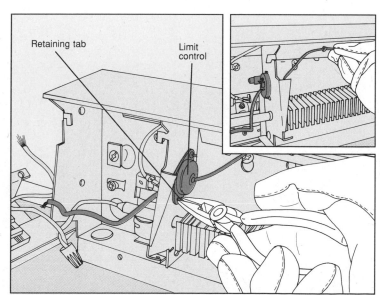

Retaining tab

Limit control

1 **Testing the limit control.** Turn off power to the baseboard heater circuit at the main service panel *(page 134)*, and test to make sure it is off *(page 85)*. Remove the control box panel *(page 86)* to access the limit control. With long-nose pliers, remove one limit control lead from its terminal. Set a multitester to the RX1K setting to test continuity *(page 136)*, then touch a probe to each limit control terminal. If there is no continuity, replace the limit control *(step 2)*.

2 **Replacing the limit control.** With power off, remove the heating element panel *(page 86)* to access the thermal conductor. Disconnect the second limit control lead. Free the limit control by unscrewing any retaining screws; if it is secured to the end panel by retaining tabs, bend out the tabs with long-nose pliers *(above)*. Also bend out the retaining tabs holding the thermal conductor. Gently work the thermal conductor *(inset)* and limit control out of the heater. Install an identical replacement in reverse order; reassemble the unit and restore the power.

SERVICING THE HEATING ELEMENT

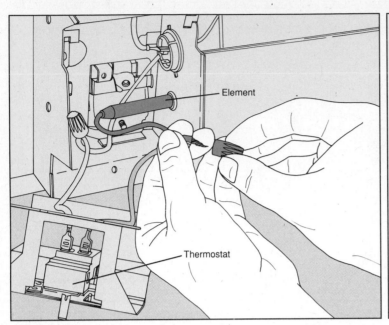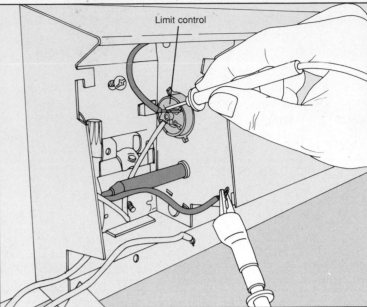

1 **Testing the element.** Turn off power to the baseboard heater circuit at the main service panel *(page 134)*, and test to make sure it is off *(page 85)*. Remove the control box panel *(page 86)*. Unscrew the wire cap connecting an element wire to a thermostat wire *(above, left)*; separate the two wires. Set a multitester to the RX1K setting to test continuity *(page 136)*. Attach a multitester alligator clip to the detached element wire, and touch the other multitester probe to the end of the wire on the limit control that leads to the other end of the element *(above, right)*. If there is no continuity, replace the heating element *(next step)* or the entire baseboard heater *(page 89)*.

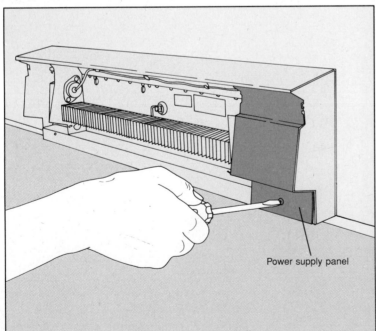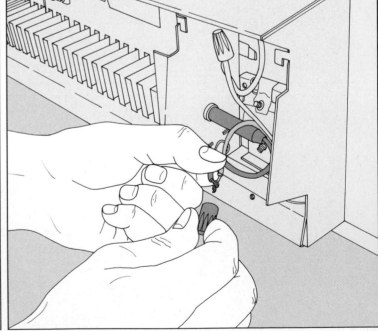

2 **Freeing the element.** With power off, remove the element panel *(page 86)* and the power supply panel *(above, left)*. Unscrew the wire cap connecting the heating element lead to the household power supply or external thermostat *(above, right)*. Leave all other wiring intact. If bumpers prevent removal of the element, lift them out of the way.

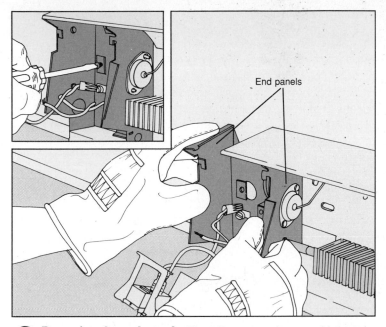

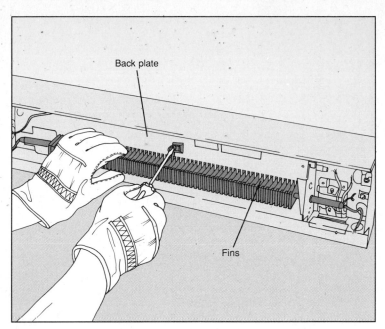

3 **Removing the end panels.** The element ends extend beyond the end panels that hold them in place. Remove any screws securing one end panel *(inset)*. Wearing gloves to protect your hands from sharp edges, slide the panel sideways *(above)*. On some models, the end panel at the other end of the unit must also be detached in order to remove the element.

4 **Removing and replacing the element.** Unscrew all retaining screws that secure the element to the back plate of the heater. Wearing gloves, hold the element gently around its fins and pull it out of the end panels. Install an identical replacement element in reverse order, making sure all wire connections are secure. Reassemble the heater and restore the power.

REPLACING A BASEBOARD HEATER

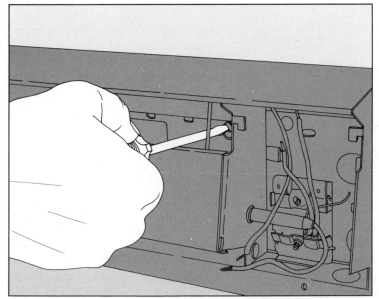

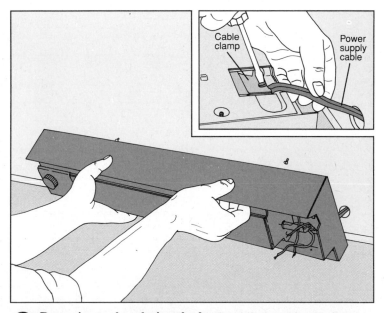

1 **Disconnecting the heater.** Turn off power to the baseboard heater circuit at the main service panel *(page 134)*, and test to make sure it is off *(page 85)*. Detach all the wires entering the unit through the back plate or end panel by unscrewing their wire caps; if the connections are permanent, cut the wires as close to the connection as possible, then strip the wires for reconnection with wire caps *(page 138)*. Unscrew and remove the ground wire. Loosen or remove the unit mounting screws on the back plate *(above)*.

2 **Removing and replacing the heater.** Lift the old unit off the retaining screws and pull it a few inches away from the wall *(above)*. Loosen the cable clamp screw *(inset)* and pull the power supply cable out of the unit. Install a new heater, reversing the steps taken to remove the old one. Connect the wires securely, assemble the unit and restore the power.

HEAT PUMPS

Instead of creating heat as conventional systems do, the heat pump uses refrigerant to transfer heat. Because the refrigerant can flow in either direction, the heat pump can be used for both heating and cooling the home. In cold weather, refrigerant in the outdoor coil, or heat exchanger, absorbs heat stored in the air. The compressor pumps the refrigerant, in hot vapor form, to the indoor coil. The blower circulates indoor air over the heated coil, warming the air for distribution through the home. As the hot vapor cools, it condenses into liquid form and flows back to the outdoor coil. During the summer, the cycle reverses; heat is absorbed from inside the home and transferred outdoors, much like central air conditioning.

Heat pumps operate most efficiently in temperatures above freezing. In mild climates, they can be used as the sole source of home heating. However, in areas where the temperature often drops below 32°F, an auxiliary heating system may be needed; electric heating is the system most commonly used.

Heat pump specialists claim that more than half of their service calls could be prevented if heat pumps were properly maintained. At the beginning of each heating season, clean the outdoor coils and straighten bent fins *(page 92)*. Check the level of the unit, and lubricate the motor and fan *(page 93)*. Before beginning most repairs or inspections, turn off power to the heat pump at the main service panel *(page 134)* and outdoor-unit disconnect switch *(page 135)*. Leave the unit off at least five minutes before switching it back on, or excess pressure could overload the compressor. **Caution:** Before working inside the unit, make a capacitor discharger *(page 140)* and discharge all capacitors *(page 94)*.

Heating or cooling problems may be caused by another unit in the system. Also consult the Troubleshooting Guides in System Controls *(page 16)* and Air Distribution *(page 24)*.

TROUBLESHOOTING GUIDE

SYMPTOM	POSSIBLE CAUSE	PROCEDURE
Heat pump doesn't run at all	No power to unit	Replace fuse or reset circuit breaker *(p. 134)* □○
	Compressor pump overloaded	Wait 30 minutes, then press reset button only once; if heat pump doesn't start, call for service
	Wiring loose or faulty	Check for loose wiring connections at the control box *(p. 96)* □◗
	Compressor contactor coil faulty	Service compressor contactor coil *(p. 99)* ■◗▲
	Compressor contactor contacts dirty	Clean contacts *(p. 99)* ◳◗
	Capacitor faulty	Discharge and test capacitors *(p. 94)* ◳◗▲ ; replace if necessary
Heat pump runs but doesn't heat or cool	Outdoor coils dirty	Clean coils *(p. 92)* □○
	Fan dirty or faulty	Lubricate fan motor *(p. 93)* □◗ ; check motor *(p. 96)* ■◗▲
	Refrigerant leaking	Call for service
Heat pump cools but doesn't heat	Indoor thermostat set to COOL	Set thermostat to HEAT
	Reversing valve stuck or faulty	Check defrost system *(p. 98)* □○ ; service reversing valve solenoid *(p. 101)* ◳◗▲ ; call for service
Heat pump heats but doesn't cool	Indoor thermostat set to HEAT	Set thermostat to COOL
	Refrigerant low	Call for service
	Reversing valve stuck or faulty	Check defrost system *(p. 98)* □○ ; service reversing valve solenoid *(p. 101)* ◳◗▲ ; call for service
Heat pump fan doesn't run	Wiring loose or faulty	Check for loose wiring connections in the electrical control box *(p. 96)* □◗
	Fan motor burned out	Test motor *(p. 96)* ◳◗▲ ; replace if necessary
	Capacitor faulty	Discharge and test capacitors *(p. 94)* ◳◗▲ ; replace if necessary
	Obstruction in fan blades	Access blades *(p. 93)* □○ ; remove obstruction
Heat pump doesn't defrost automatically; ice buildup on coils	Indoor thermostat set to HEAT	Set thermostat to AUTOMATIC HEAT
	Air-flow sensor tube blocked	Clear air-flow sensor tube *(p. 93)* □◗
	Coil fins flattened	Clean and straighten coil fins *(p. 92)* □◗▲
	Reversing valve stuck or faulty, or defrost system faulty	Check defrost system *(p. 98)* □○ ; service reversing valve solenoid *(p. 101)* ◳◗▲ ; call for service
Auxiliary light on indoor thermostat lit constantly	Reversing valve stuck or faulty	Check defrost system *(p. 98)* □○ ; service reversing valve solenoid *(p. 101)* ◳◗▲ ; call for service
	Outdoor temperature sensor faulty	Test sensor *(p. 98)* ◳◗▲ ; replace if necessary

DEGREE OF DIFFICULTY: □ Easy ◳ Moderate ■ Complex
ESTIMATED TIME: ○ Less than 1 hour ◗ 1 to 3 hours ● Over 3 hours   ▲ Special tool required

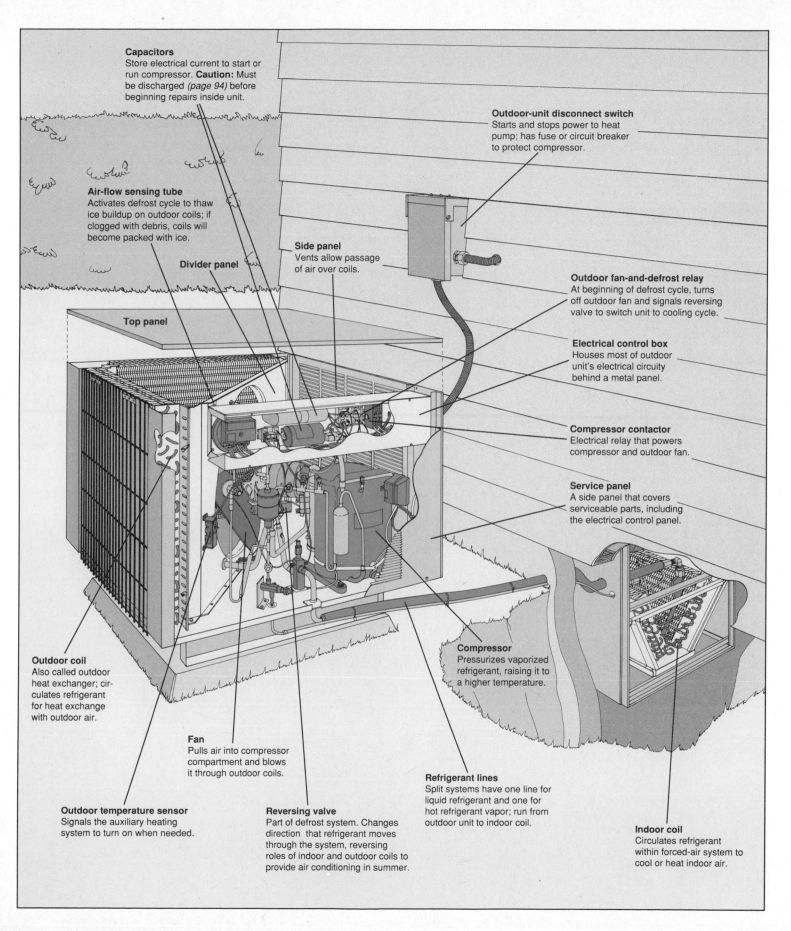

Capacitors
Store electrical current to start or run compressor. **Caution:** Must be discharged *(page 94)* before beginning repairs inside unit.

Air-flow sensing tube
Activates defrost cycle to thaw ice buildup on outdoor coils; if clogged with debris, coils will become packed with ice.

Divider panel

Top panel

Side panel
Vents allow passage of air over coils.

Outdoor-unit disconnect switch
Starts and stops power to heat pump; has fuse or circuit breaker to protect compressor.

Outdoor fan-and-defrost relay
At beginning of defrost cycle, turns off outdoor fan and signals reversing valve to switch unit to cooling cycle.

Electrical control box
Houses most of outdoor unit's electrical circuity behind a metal panel.

Compressor contactor
Electrical relay that powers compressor and outdoor fan.

Service panel
A side panel that covers serviceable parts, including the electrical control panel.

Outdoor coil
Also called outdoor heat exchanger; circulates refrigerant for heat exchange with outdoor air.

Fan
Pulls air into compressor compartment and blows it through outdoor coils.

Compressor
Pressurizes vaporized refrigerant, raising it to a higher temperature.

Outdoor temperature sensor
Signals the auxiliary heating system to turn on when needed.

Reversing valve
Part of defrost system. Changes direction that refrigerant moves through the system, reversing roles of indoor and outdoor coils to provide air conditioning in summer.

Refrigerant lines
Split systems have one line for liquid refrigerant and one for hot refrigerant vapor; run from outdoor unit to indoor coil.

Indoor coil
Circulates refrigerant within forced-air system to cool or heat indoor air.

ACCESS TO THE INTERNAL COMPONENTS

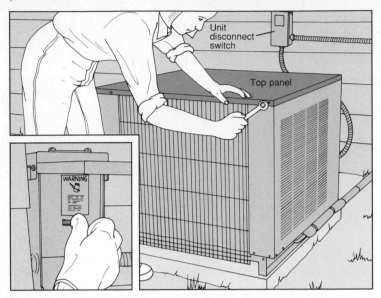

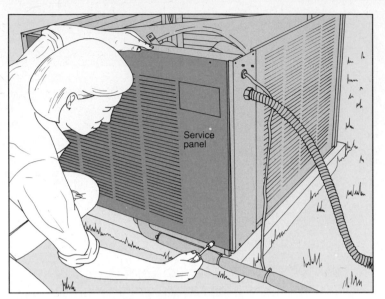

1 **Removing the top panel.** Turn off power to the heat pump at the main service panel *(page 134)* and at the outdoor-unit disconnect switch *(inset)*. Use a socket wrench or nut driver to remove the sheet metal screws securing the top panel around its edges *(above)*. If the screws are stubborn, loosen them with a few drops of penetrating oil. Wearing work gloves, lift off the panel; place it clear of the unit and refrigerant lines.

2 **Removing the side panels.** The service panel, usually identified by an information sticker, must be removed for access to the compressor and electrical control box. The other side panels must be removed for cleaning the evaporator coils. With power to the heat pump switched off *(step 1)*, use a nut driver to remove the sheet metal screws securing the side panels *(above)*. Wearing work gloves, pull the panels away from the unit and place them clear of the refrigerant lines.

MAINTAINING THE HEAT PUMP

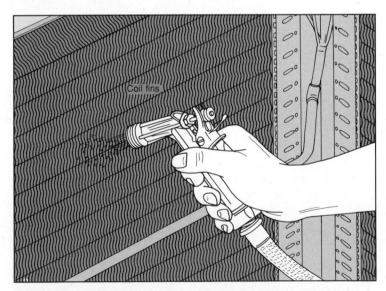

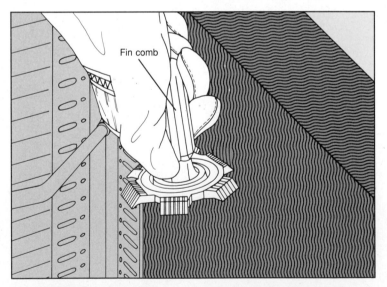

Cleaning the coils. With power to the heat pump turned off, remove the top and side panels *(steps above)* to access the coils. Use a soft brush and a portable cordless vacuum cleaner to clean the coil fins. To remove stubborn dirt buildup, spray water through the coil fins from inside, using a high-pressure nozzle on a garden hose *(above)*. **Caution:** Do not use a knife or screwdriver blade to dislodge dirt between fins; these tools can puncture the coils. Sweep debris accumulated on the bottom of the heat pump into a dustpan. Use a fin comb to straighten any bent coil fins *(right)*.

Straightening coil fins. With power to the heat pump turned off, remove the top and side panels *(steps above)* to access the coils. Wearing gloves, use a multi-head fin comb *(page 132)* to straighten bent coil fins. Determine which head of the comb corresponds to the spacing of fins on the coils; the teeth on the head should fit easily between the fins. Gently fit the teeth of the fin comb between the coil fins in an undamaged section below the area to be straightened. Pull the fin comb up, sliding it through the damaged area *(above)*; at the same time, comb out any lodged debris. **Caution:** Do not use a knife or screwdriver blade to straighten fins; the coils could be damaged.

MAINTAINING THE HEAT PUMP (continued)

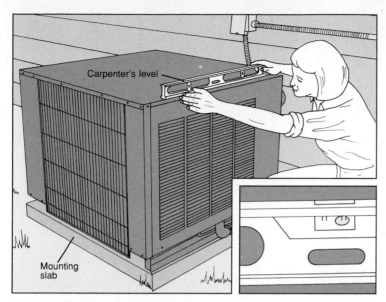

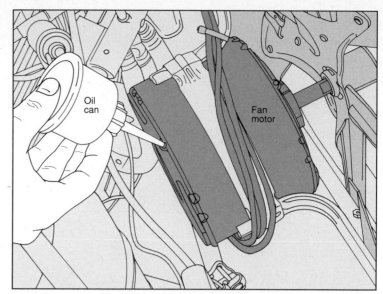

Checking the slope for proper drainage. If your heat pump is seated on a mounting slab, as shown above, check the slab's level each spring. Place a carpenter's level on top of the unit *(above)*. The level should read one-half to one bubble off center *(inset)*, sloping away from the house. Otherwise, the mounting slab has settled. Adjust the slab to its proper slope: Use a pry bar to lift the slab and slide a brick or other prop underneath, or call a professional to adjust it with concrete.

Lubricating the fan motor. Turn off power to the heat pump at the main service panel *(page 134)* and outdoor-unit disconnect switch *(page 135)*. Remove the top panel *(page 92)*. Pry off the plastic or metal oil-port caps. Squirt a few drops of SAE 10 non-detergent oil into the oil ports *(above)*. If the fan motor is sealed and has no oil ports, squirt a little oil on the fan shaft where it meets the motor. Turn the fan blade back and forth a few times to work in the oil. Check for obstructions in the blade and remove them.

CLEARING THE AIR-FLOW SENSING TUBE

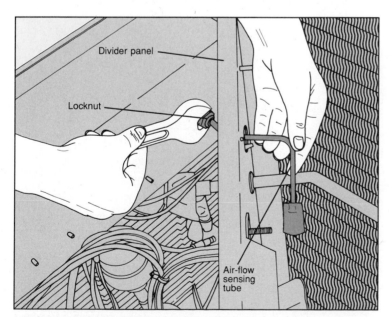

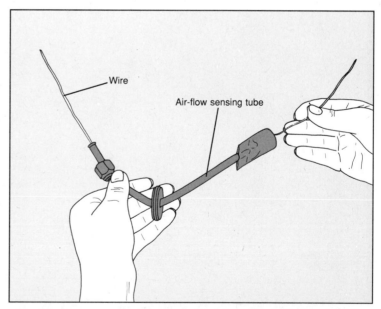

1 **Removing the air-flow sensing tube.** Turn off power to the heat pump at the service panel *(page 134)* and outdoor unit disconnect switch *(page 135)*. Remove the top panel *(page 92)*, and locate the air-flow sensing tube, a short, narrow pipe mounted on the divider panel. Use an open-end wrench to loosen the tube's locknut, and work the tube out of its opening in the divider panel *(above)*.

2 **Clearing the sensing tube.** To clear a debris-filled sensing tube, feed a flexible wire or pipe cleaner through the opening in the tube, sliding it in and out until the tube is cleaned *(above)*. Next, pull out the wire and test the tube by blowing air through it; the air should flow freely. Reinstall the tube, threading it back through its opening in the divider panel; tighten the locknut. Reassemble the heat pump and turn on the power.

DISCHARGING THE CAPACITORS

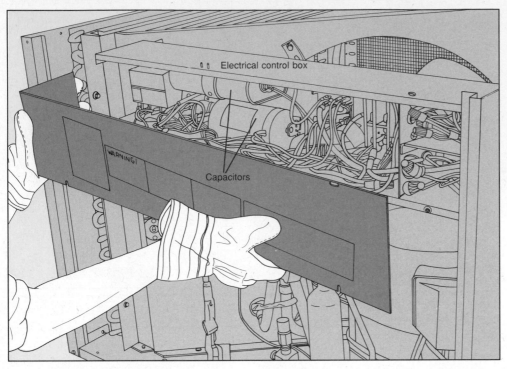

Electrical control box

Capacitors

1 **Removing the electrical control-box panel. Caution:** Capacitors may store potentially dangerous voltage. Before servicing the control box components *(next step)*, make a capacitor discharging tool *(page 140)* and discharge the heat pump capacitors. Turn off power to the heat pump at the main service panel *(page 134)* and outdoor-unit disconnect switch *(page 135)*. Access the internal components *(page 92)*. Use a socket wrench or nut driver to remove the sheet metal screws holding the electrical control-box panel in place. Wearing work gloves to protect your hands, pull off the panel *(left)*.

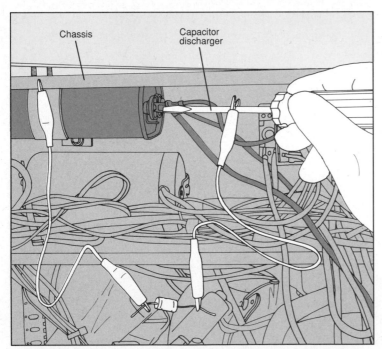

Chassis

Capacitor discharger

2 **Discharging an uncapped capacitor.** Make a capacitor discharger *(page 140)*. With power to the heat pump turned off, clip one end of the capacitor discharger to the heat pump chassis. If a capacitor has more than two terminals and no bleed resistor, discharge it as shown on page 141. If it has a bleed resistor, go to steps 3 and 4. If a capacitor has two terminals and no bleed resistor, as shown above, hold the insulated screwdriver handle of the capacitor discharger in one hand and touch each terminal, one at a time, with the tip of the screwdriver blade for one second *(above)*. Discharge all other capacitors in the unit.

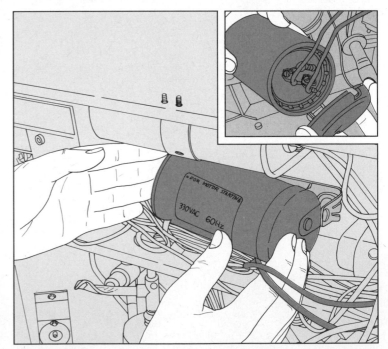

3 **Removing a capped capacitor.** With power to the heat pump turned off, place one hand at each end of the capacitor, holding the cap on, and pull the capacitor out of its mounting bracket in the electrical control box *(above)*. Lift off the cap to expose the capacitor wire terminals *(inset)*. **Caution:** Do not touch the terminals with your hands.

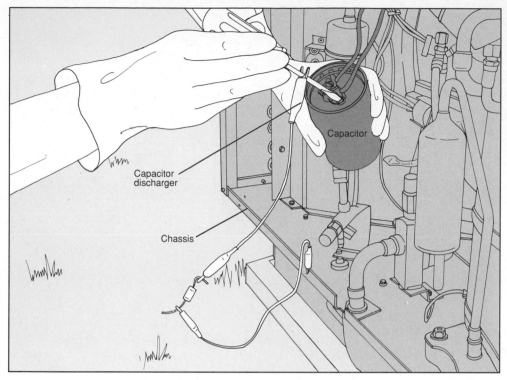

Capacitor

Capacitor discharger

Chassis

4 **Discharging a capped capacitor.** With power to the heat pump turned off, clip one end of the capacitor discharger to the heat pump chassis. If a capacitor has no bleed resistor and more than two terminals, discharge it as shown on page 141. If a capacitor has two terminals and a bleed resistor, as shown at left, hold the insulated screwdriver handle of the capacitor discharger in one hand. Touch each terminal, one at a time, with the tip of the screwdriver blade for one second. Discharge all other capacitors in the heat pump.

TESTING AND REPLACING THE CAPACITORS

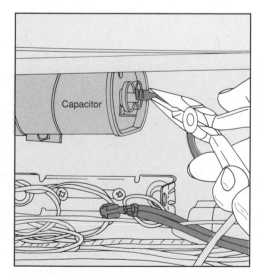

Capacitor

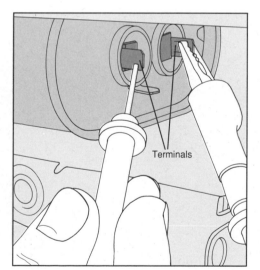

Terminals

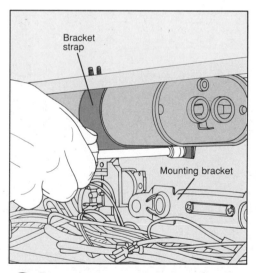

Bracket strap

Mounting bracket

1 **Inspecting the capacitor casing.** Turn off power to the heat pump at the main service panel *(page 134)* and outdoor-unit disconnect switch *(page 135)*. Discharge the capacitors *(page 94)*. Visually check each capacitor. If you see a bulging, broken or melted casing, replace the capacitor *(step 3)*. If the casing looks OK, use insulated long-nose pliers to slip the wire leads off the first capacitor you are testing *(above)*.

2 **Testing the capacitor.** Make sure power to the heat pump is off, and set a multitester to the RX1K scale. Test a three-terminal capacitor as shown on page 125. Test a two-terminal capacitor by placing one multitester probe on each of the capacitor's terminals *(above)* while watching the tester needle. The needle should swing to zero resistance, then slowly move a third to halfway across the scale toward infinity. If the needle swings to zero resistance and stays there, or if there is no movement of the needle, the capacitor is faulty; replace it *(next step)*.

3 **Removing and replacing the capacitor.** A capped capacitor snaps out of a flexible mounting bracket. An uncapped capacitor is usually mounted with bracket straps to the back panel of the electrical control box; use a nut driver to remove the sheet metal screw holding the bracket in place *(above)*. **Caution:** Keep the capacitor out of the reach of children and do not incinerate. Replace it with an identically rated one.

CHECKING AND SERVICING WIRING CONNECTIONS

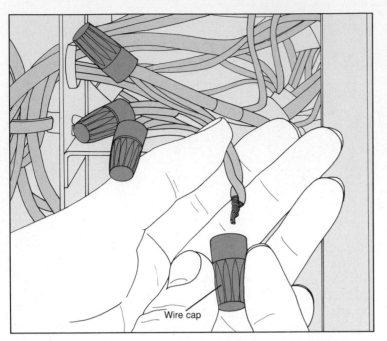

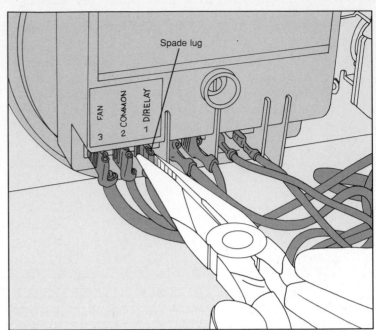

1 **Checking wire cap connections.** Turn off power to the heat pump at the main service panel *(page 134)* and outdoor-unit disconnect switch *(page 135)*. Access the electrical control box and discharge the capacitors *(page 94)*. Twist off each wire cap and check the wire leads. If the wires have separated, twist them together clockwise and screw on the wire cap *(above)*.

2 **Checking spade lug connectors and wires.** With power to the unit off and capacitors discharged, inspect the electrical wiring inside the electrical control box. Wires with frayed or charred insulation should be replaced; call a professional. Also inspect all spade lug connections: Using insulated long-nose pliers, push spade lugs securely onto their terminals *(above)*.

SERVICING THE FAN MOTOR

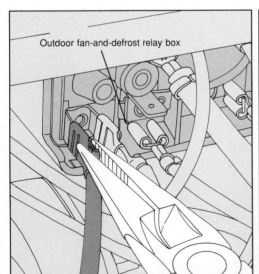

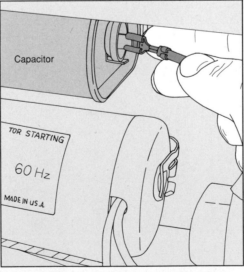

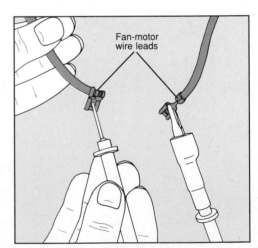

1 **Disconnecting the wire leads.** Turn off power to the heat pump at the main service panel *(page 134)* and outdoor-unit disconnect switch *(page 135)*. Access the electrical control box and discharge all capacitors *(page 94)*. Trace the fan motor wires from the motor through an inlet hole in the back panel of the electrical control box. Locate the fan motor wire to the outdoor fan-and-defrost relay box and the fan motor wire to the compressor run capacitor. Label these wires for correct reconnection *(page 135)*. Using insulated long-nose pliers, disconnect the fan-motor wire lead from the relay box *(above, left)*, then disconnect the wire lead from the capacitor *(above, right)*.

2 **Testing the fan motor.** With power to the unit turned off, set a multi-tester to test continuity *(page 136)* and touch a probe to each disconnected wire lead *(above)*. If the multitester shows continuity and you do not smell an electrical odor, the fan motor is OK; reconnect the leads. If there is continuity and an electrical odor, the fan motor is shorted; if there is no continuity, the motor has an open circuit. In either case, replace the fan motor *(next step)*.

SERVICING THE FAN MOTOR (continued)

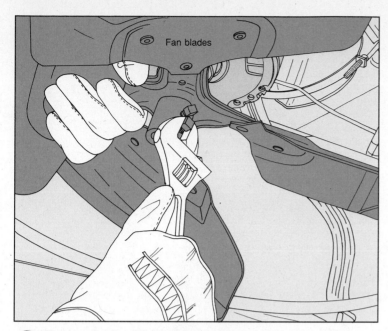

Fan blades

3 **Freeing the fan blades.** Wear work gloves to protect your hands from the sharp blades. With power to the unit turned off, hold one of the fan blades steady and use an adjustable wrench to loosen the two square-head bolts *(above)* or screws that secure the fan blades to the motor shaft.

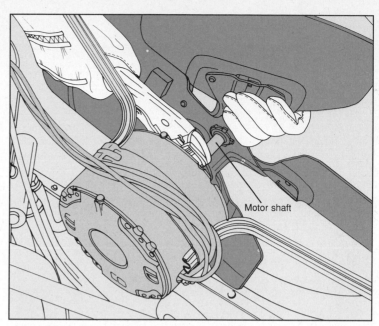

Motor shaft

4 **Dismounting the fan blades.** Wearing work gloves, use a pair of locking pliers to hold the motor shaft. Being careful not to bend the blades, twist the fan back and forth on the motor shaft *(above)*. If the fan is stuck or rusted, loosen it by squirting penetrating oil on the junction of the motor shaft and fan blades. If the fan blades still won't budge, apply more penetrating oil, wait a few minutes and try again. Once the blades are free, pull them straight off the motor shaft; place them face down, away from the unit and refrigerant lines.

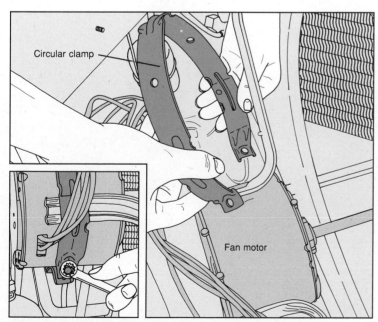

Circular clamp

Fan motor

5 **Removing the fan mounting clamp.** A flexible circular clamp keeps the fan motor firmly in place between three bracket arms. Use a combination wrench—or two if necessary—to loosen the bolt and nut that join the two ends of the clamp *(inset)*. Then open the clamp just enough to slip it off the motor and bracket arms *(above)*. Avoid dropping the heavy fan motor by using one hand to grasp the motor shaft and cupping the motor in your other hand.

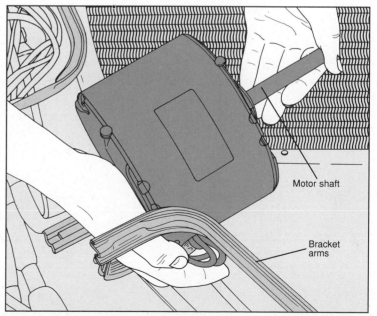

Motor shaft

Bracket arms

6 **Removing the fan motor.** Slip the fan motor straight forward out of the three bracket arms *(above)*. Label and disconnect the rest of the fan motor wires and pull them out through their inlet hole in the control box. Look for an identification label affixed to the motor casing; bring this information to the heat pump dealer and purchase an exact replacement fan motor. The replacement part comes equipped with wire leads. Install it reversing the sequence here; reassemble the heat pump, and turn on the power.

CHECKING THE DEFROST SYSTEM

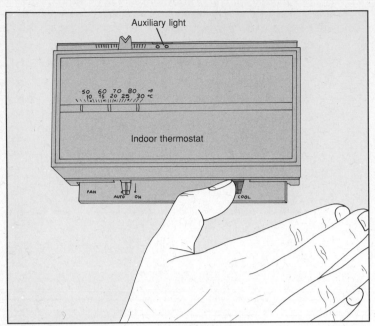

Auxiliary light

Indoor thermostat

Troubleshooting the defrost system. The auxiliary light on the indoor thermostat indicates that the heat pump is in defrost mode, or that a low outdoor temperature has activated the auxiliary heating system. If the light remains continously lit and you see ice buildup on the outdoor coils, set the indoor thermostat to COOL *(left)*. This should deliver warm refrigerant from the indoor coils to the outdoor coils, thawing the ice. Wait at least 30 minutes, then inspect the outdoor coils again. If the ice has thawed, the reversing valve may have been stuck; reset the thermostat to HEAT. If the problem recurs, either the defrost sensing system or the reversing valve is faulty; call for service. If the coils remain iced, the reversing valve solenoid may be defective; service the reversing valve *(page 101)*.

SERVICING THE OUTDOOR TEMPERATURE SENSOR

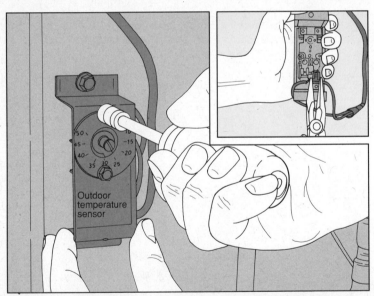

Outdoor temperature sensor

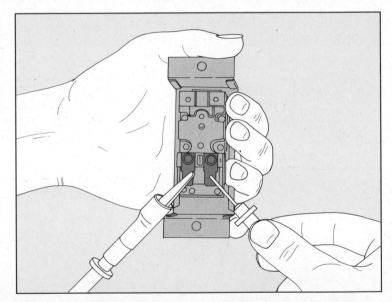

1 **Removing the temperature sensor.** If the auxiliary light on the indoor thermostat remains lit constantly, the outdoor temperature sensor—if your heat pump has one—may be faulty. Turn off power to the heat pump at the main service panel *(page 134)* and outdoor-unit disconnect switch *(page 135)*. Access the internal components *(page 92)*, and discharge all capacitors *(page 94)*. Locate the outdoor temperature sensor, usually mounted in the compressor compartment as shown above, but sometimes outside the heat pump in a mounting box. Use a nut driver to undo the screws that mount the sensor *(above)*. Label the wires, then use long-nose pliers to disconnect the sensor wire connectors from their terminals *(inset)*. Since the test shown in step 2 will work only at low temperatures, place the temperature sensor in the freezer for one-half hour.

2 **Testing the temperature sensor for continuity.** Remove the sensor from the freezer. Set a multitester to test continuity *(page 136)*, then touch a probe to each sensor terminal *(above)*. If the multitester shows continuity, the temperature sensor is working; if there is no continuity, replace the sensor with an identical part, reassemble the heat pump and turn on the power.

SERVICING THE COMPRESSOR CONTACTOR

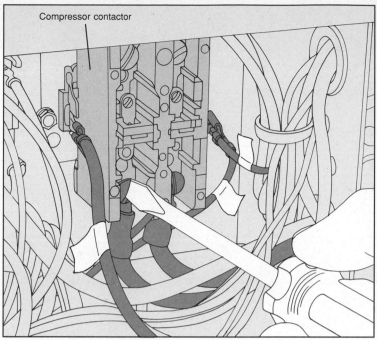

Compressor contactor

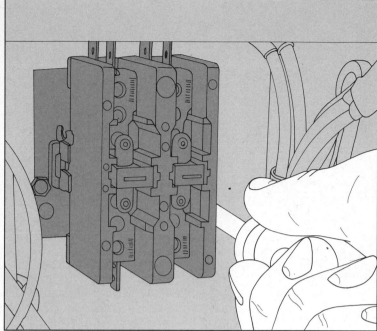

1 **Disconnecting the leads.** Turn off power to the heat pump at the main service panel *(page 134)* or outdoor-unit disconnect switch *(page 135)*. Access the internal components *(page 92)* and discharge the capacitors *(page 94)*. Label the wires for correct reconnection *(page 135)*, then pull off the connectors, or use a screwdriver to remove the screws that attach the wire leads to terminals on the contactor *(above)*. Use a nut driver to remove the screws that mount the contactor to the control panel *(above, right)*, and lift it out of the control box.

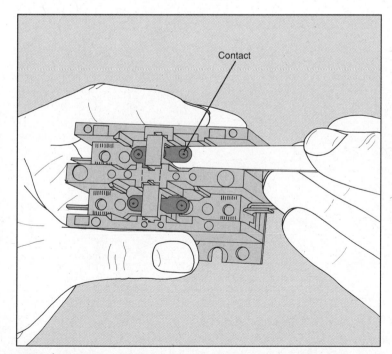

Contact

2 **Cleaning the contacts.** Visually inspect the reed-style contacts. If the contacts are stuck together, replace the contactor with an identical part *(step 3)*. Using an ice cream stick, rub off any dirt buildup or oxidation *(above)*. If the contacts look burned or pitted, insert a piece of sandpaper between them and slide it back and forth. If the contacts cannot be cleaned, replace the contactor. If the contactor looks OK, test it.

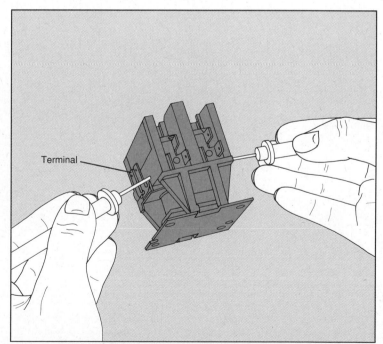

Terminal

3 **Testing the compressor contactor coil.** Set a multitester to test continuity *(page 136)*, then touch a probe to each coil terminal on the sides of the contactor *(above)*. If there is continuity, the coil is working; reinstall it. If there is no continuity, replace the contactor with an identical part. Reassemble the heat pump and turn on the power.

TESTING THE COMPRESSOR

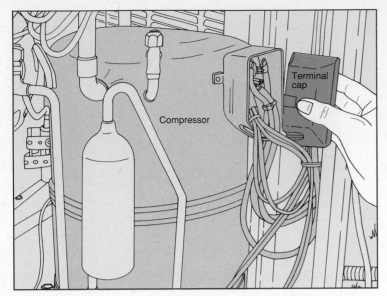

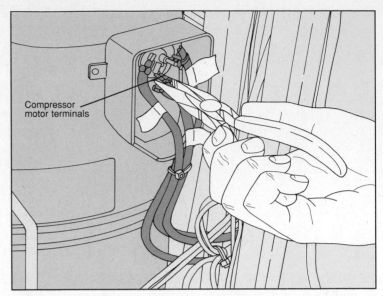

1 **Removing the compressor terminal cap.** Turn off power to the heat pump at the main service panel *(page 134)* and the outdoor-unit disconnect switch *(page 135)*. Access the electrical control box *(page 92)* and discharge the capacitors *(page 94)*. Locate the compressor terminal cap, a plastic cover that protects the compressor motor terminals, with an inlet hole for the motor leads. Grasp the cap and pull it straight off *(above)*.

2 **Disconnecting the compressor leads.** Since each compressor manufacturer uses a different sequence of terminals (R for run winding, S for start winding, C for common), it is important to label each wire before removing it *(page 141)*. If the terminals are not marked, draw a diagram of the wire lead positions on the compressor motor terminals. Use long-nose pliers to pull the lead connectors straight off their terminals *(above)*.

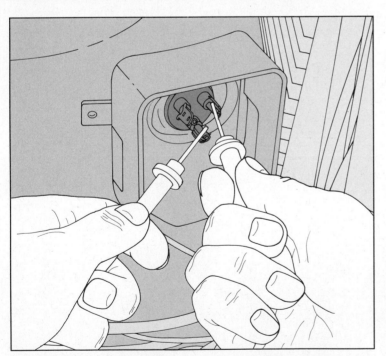

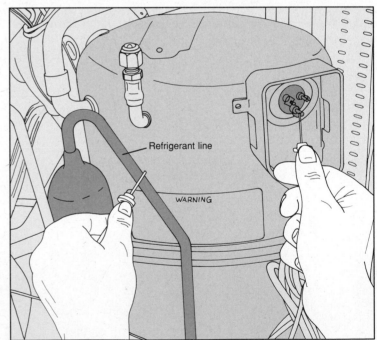

3 **Testing the compressor motor windings.** Set a multitester to test continuity *(page 136)*. Touch one probe to one compressor motor terminal and touch the other probe to another terminal *(above)*. Repeat the test for all three possible combinations of terminals; there should be continuity each time. If there is continuity, the compressor motor windings are working; go to step 4. If there is no continuity in any one of the tests, the compressor must be replaced; call for service.

4 **Testing the compressor motor for ground.** To ensure a correct reading, allow the compressor motor to cool for three to four hours before conducting this test. Set a multitester to test continuity *(page 136)*. Touch one probe to one of the compressor motor terminals, and touch the other probe to a clean, unpainted metal surface on the compressor dome or to one of the copper refrigerant lines leading from the compressor *(above)*. Repeat the test for the other two compressor terminals. If all tests show no continuity, the compressor motor windings are OK. If any test shows continuity, the windings are grounded, and the compressor must be replaced; call for service.

SERVICING THE REVERSING VALVE SOLENOID

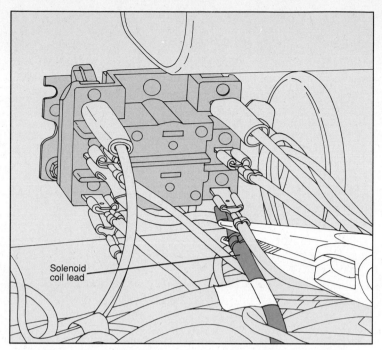

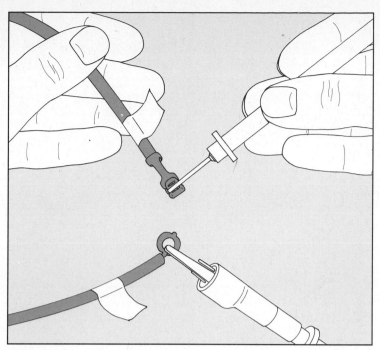

Solenoid
coil lead

1 **Disconnecting the leads.** Turn off power to the heat pump at the main service panel *(page 134)* and outdoor-unit disconnect switch *(page 135)*. Access the compressor compartment *(page 92)* and the electrical control box and discharge all capacitors *(page 94)*. Locate the solenoid coil and trace the two wire leads through the inlet opening in the electrical control box. One goes to the fan-and-defrost control relay box; the other is screwed to the chassis. Label the leads for correct reconnection, then disconnect them *(above)*.

2 **Testing the solenoid coil.** Set a multitester to test continuity, then touch a tester probe to each solenoid coil lead *(above)*. If the multitester indicates continuity, the coil is OK; check for a stuck reversing valve *(page 98)* or faulty wiring connections *(page 96)*. If there is no continuity, the coil has an open circuit; remove and replace it *(step 3)*.

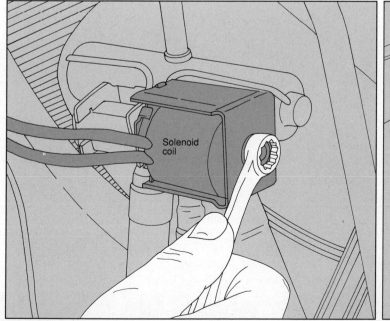

Solenoid
coil

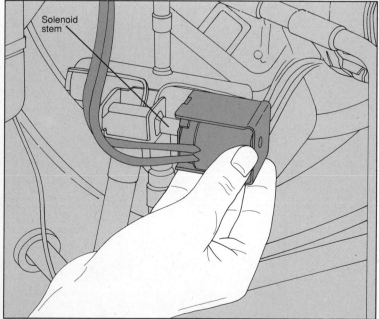

Solenoid
stem

3 **Removing and replacing the solenoid coil.** Use a box-end wrench to remove the solenoid's locknut *(above, left)*. Slip the solenoid coil off its stem. Slip an identical replacement coil—which comes with leads—onto the solenoid stem *(above, right)*. Tighten the locknut with a box-end wrench. Thread the solenoid leads through the inlet opening in the control box. Use long-nose pliers and a nut driver or screwdriver to connect the leads to the correct terminals.

CENTRAL AIR CONDITIONING

In addition to cooling the air, central air conditioning also dehumidifies and circulates it. The system consists of two parts: the condenser unit, located outdoors, and the evaporator coil, mounted inside the furnace, usually in the plenum.

Central air conditioners are typically part of a forced-air heating system, allowing both heating and cooling to share the ducting and blower. In some cases, an integrated system located in either the attic or the basement contains both the condenser and evaporator coils, and has its own ducting.

When the thermostat calls for cooling, it switches on the air conditioner, which cools liquid refrigerant in the condenser coils. The compressor sends cooled refrigerant through one of two refrigerant lines to the evaporator coils inside the house. When the cooled refrigerant reaches the indoor evaporator coils, the furnace blower circulates warm air from the house over the cold coils. The evaporator coils absorb heat from the indoor air as the liquid refrigerant is transformed into a gaseous state. This gas is pumped outdoors through a second refrigerant line to the condenser coils. There the heat that was absorbed indoors is released.

At the beginning of each cooling season, clean outdoor condenser coils, and straighten the fins if necessary *(page 108)*.

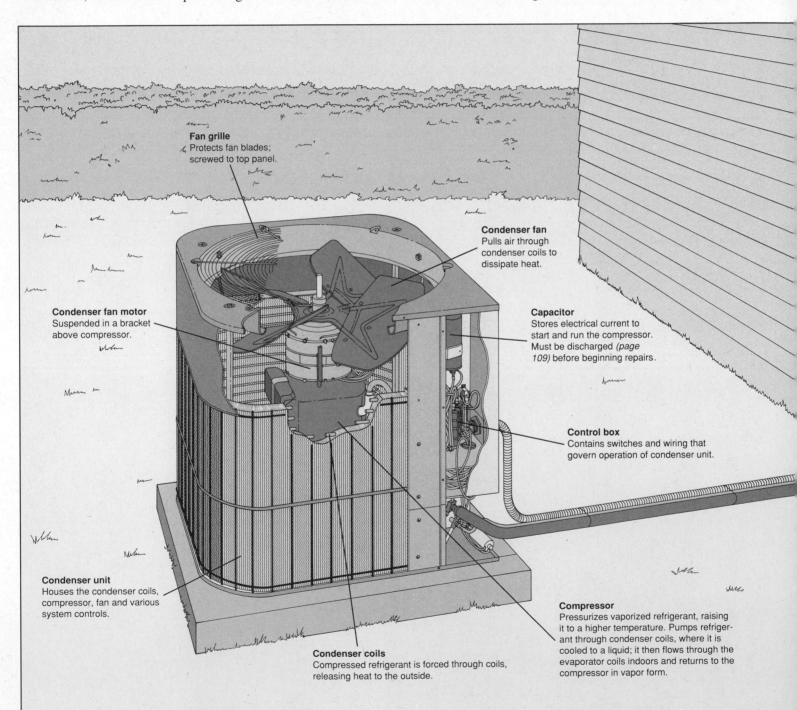

Fan grille
Protects fan blades; screwed to top panel.

Condenser fan
Pulls air through condenser coils to dissipate heat.

Condenser fan motor
Suspended in a bracket above compressor.

Capacitor
Stores electrical current to start and run the compressor. Must be discharged *(page 109)* before beginning repairs.

Control box
Contains switches and wiring that govern operation of condenser unit.

Condenser unit
Houses the condenser coils, compressor, fan and various system controls.

Condenser coils
Compressed refrigerant is forced through coils, releasing heat to the outside.

Compressor
Pressurizes vaporized refrigerant, raising it to a higher temperature. Pumps refrigerant through condenser coils, where it is cooled to a liquid; it then flows through the evaporator coils indoors and returns to the compressor in vapor form.

Check the level of the outdoor condenser unit to ensure correct refrigerant flow between indoor and outdoor coils *(page 107)*. Lubricate the motor *(page 107)* and make sure the air conditioner fan rotates properly *(page 109)*.

Cold evaporator coils condense moisture out of the indoor air. This water drips into a drain pan beneath the coils. Clean the drain pan under a V-shaped evaporator coil, and clear the condensate drain pipe running from the drain pan under an A-shaped coil, to keep water from puddling inside the furnace and to prevent the growth of algae and bacteria. Before beginning a repair, turn off power to the air conditioner at the main

service panel *(page 134)* and outdoor-unit disconnect switch *(page 135)*. Leave the unit off for at least five minutes before turning it back on to prevent excess refrigerant pressure from overloading the compressor. **Caution:** Discharge all capacitors *(page 109)* by making a capacitor discharger *(page 140)* for repairs that advise it. Avoid handling fins, coils or refrigerant lines; these carry high-pressure refrigerant and should be serviced only by a professional.

Cooling problems may be caused by another unit in the system. Consult the Troubleshooting Guides in System Controls *(page 16)* and Air Distribution *(page 24)*.

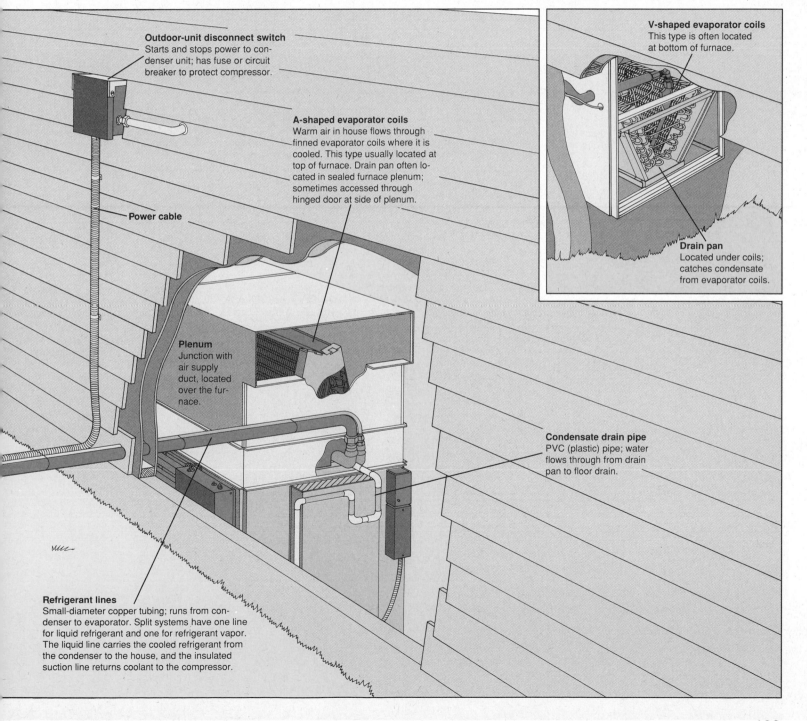

V-shaped evaporator coils
This type is often located at bottom of furnace.

Drain pan
Located under coils; catches condensate from evaporator coils.

Outdoor-unit disconnect switch
Starts and stops power to condenser unit; has fuse or circuit breaker to protect compressor.

A-shaped evaporator coils
Warm air in house flows through finned evaporator coils where it is cooled. This type usually located at top of furnace. Drain pan often located in sealed furnace plenum; sometimes accessed through hinged door at side of plenum.

Power cable

Plenum
Junction with air supply duct, located over the furnace.

Condensate drain pipe
PVC (plastic) pipe; water flows through from drain pan to floor drain.

Refrigerant lines
Small-diameter copper tubing; runs from condenser to evaporator. Split systems have one line for liquid refrigerant and one for refrigerant vapor. The liquid line carries the cooled refrigerant from the condenser to the house, and the insulated suction line returns coolant to the compressor.

TROUBLESHOOTING GUIDE

SYMPTOM	POSSIBLE CAUSE	PROCEDURE
Condenser unit does not turn on	No power to unit	Turn on outdoor-unit disconnect switch; replace fuse or reset circuit breaker *(p. 134)* □○
	Thermostat set too high	Lower the thermostat setting
	Outdoor temperature sensor faulty	Test temperature sensor and replace if necessary *(p. 112)* ▬●▲
	High-pressure switch tripped	Reset high-pressure switch *(p. 113)* □○
	High-pressure switch faulty	Test switch and replace if necessary *(p. 113)* ▬●▲
	Compressor faulty	Test compressor motor *(p. 114)* ▬●▲ ; call for service
	Capacitor faulty	Discharge capacitors *(p. 109)* □○▲ ; test and replace if necessary *(p. 110)* ▬●▲
	Fan motor faulty	Test fan motor and replace if necessary *(p. 113)* ▬●▲
Condenser unit does not turn off	Contactor faulty	Clean contactor *(p. 110)* □○; test and replace if necessary *(p. 110)* ▬●▲
Condenser unit noisy	Fan blades hitting bracket or grille	Inspect fan blades; adjust or replace if necessary *(p. 109)* □●▲
	Fan motor faulty	Test fan motor; replace if necessary *(p. 113)* ▬●▲
	Access panel screws loose	Tighten screws
	Fan motor bearings dry	Lubricate fan motor *(p. 113)* □○
Air conditioning does not cool	Thermostat set too high	Lower the thermostat setting
	Fan not cooling condenser coils	Test the fan motor and replace if necessary *(p. 113)* ▬●▲
	High-pressure switch tripped	Reset high-pressure switch *(p. 113)* □○
	High-pressure switch faulty	Test switch and replace if necessary *(p. 113)* ▬●▲
	Compressor faulty	Test compressor *(p. 114)* ▬○▲ ; call for service
	Refrigerant level low	Call for service
	Condenser coils blocked with dirt or grass	Clean coils and remove obstruction *(p. 108)* □○
	Condenser unit blocked; air cannot circulate around unit	Clear debris from around unit; call for service to determine whether unit must be moved
Air conditioning short cycles (turn on and off repeatedly)	Evaporator coils frosted	Run the blower in furnace with the air conditioning turned off for several hours
	Refrigerant level low	Call for service
Frost on evaporator coils	Blower in furnace turned off	Turn on blower *(Air Distribution, p. 24)*
	Blower motor faulty	Call for service
Water leaking inside furnace	Evaporator drain trap blocked	Clean out drain trap *(p. 106)* □○
	Evaporator drain pan clogged	Clean drain pan *(p. 105)* □○

DEGREE OF DIFFICULTY: □ Easy ▬ Moderate ■ Complex
ESTIMATED TIME: ○ Less than 1 hour ● 1 to 3 hours ● Over 3 hours   ▲ Special tool required

MAINTAINING THE EVAPORATOR

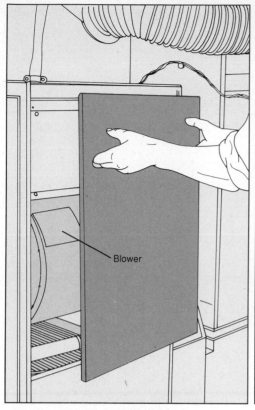

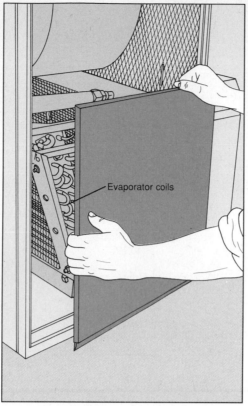

Blower

Evaporator coils

1 **Accessing the evaporator coils.** Turn off power to the air conditioner and the furnace at the main service panel *(page 134)* and outdoor-unit disconnect switch. In most furnaces, the evaporator coils are top-mounted and often not accessible: The plenum that houses the coils may be a sealed unit, or ductwork and refrigerant lines may block access. A service technician may be able to install an access panel in a sealed plenum. In the air conditioner/electric furnace combination shown here, the evaporator coils are bottom-mounted and easily reached through access panels. First remove the blower panel *(far left)* by inserting your fingers in the slots and pulling it up and out. Then lift off the coil panel *(near left)*.

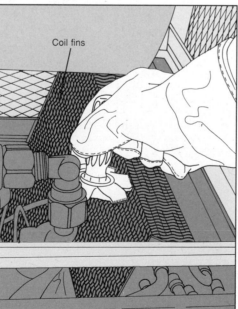

Coil fins

Drain pan

2 **Cleaning and straightening the coil fins.** Dirt that is not trapped by the air filter can lodge in the evaporator coil fins. With power to the unit off, clean the coils if necessary before each cooling season, or after running the air conditioner with a dirty filter. Using a very soft brush, gently stroke the coils up and down along the fins *(above, left)*. Brush all surfaces of the coils, paying special attention to the air-intake side (the side facing the air filter). Use a mild detergent-and-water solution for stubborn dirt; if the coils are top-mounted, do not let water drip on the heating elements or heat exchanger.

Check the coils for bent fins. Wearing work gloves, fit the teeth of a fin comb *(page 132)* into the fins near a damaged area, and gently pull the comb through the fins to unbend them *(above, right)*.

3 **Cleaning the drain pan.** Algae and sediment may collect in the drain pan. Wearing work gloves to protect your hands from the fins, wipe as much of the pan as you can reach, using a sponge. Flush out the pan and drain tube using a garden hose *(above)* or a pitcher of water. To prevent algae formation, add one-half cup bleach to the water in the drain pan. Alternatively, use algaecide pills, following the manufacturer's instructions.

SERVICING THE DRAIN TUBE

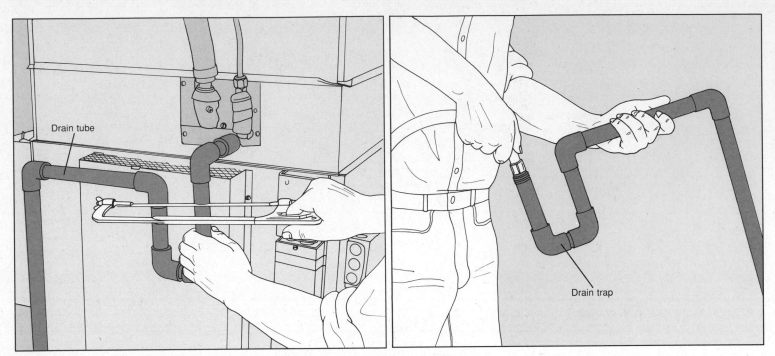

1 **Clearing a blockage in the trap.** The drain tube from the evaporator coil drain pan is usually made of PVC (plastic) pipe. The joints are permanently glued. The trap, which prevents insects and warm, humid basement air from entering the plenum, may become blocked by sediment or algae. Access the blockage by cutting the tube straight across with a hacksaw *(above, left)*, 1 to 2 inches past the angled joint leading from the plenum. Flush out the trap using a garden hose *(above, right)*. To help prevent blockage by algae, pour a solution of one tablespoon bleach and one-half cup water into the trap; alternatively, drop in an algaecide tablet.

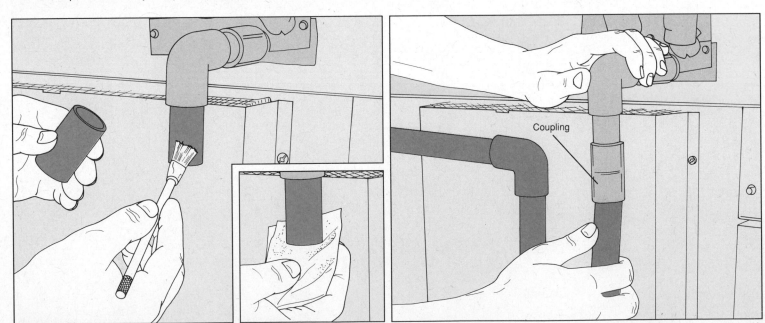

2 **Reassembling the drain tube.** Use sandpaper to deburr the rough edges of the cut ends of the tube *(inset)*. Buy a straight PVC coupling and test its size by sliding it over a tube end—it should fit snugly. Remove the coupling. Brush PVC solvent cement onto the outside surfaces of the two ends *(above, left)*, and immediately reassemble the drain tube with the coupling *(above, right)*. Twist the tube to the correct angle before the cement sets: The bottom of the drain tube should be positioned as close as is practical to the basement drain.

ACCESSING THE CONDENSER UNIT

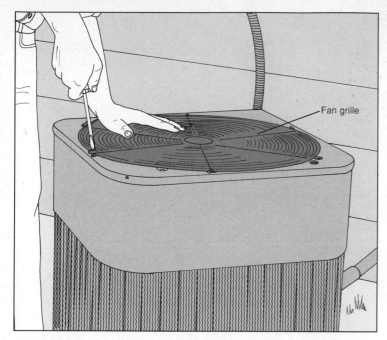

Fan grille

Removing the fan grille. Turn off power to the condenser unit at the main service panel *(page 134)* and outdoor-unit disconnect switch. Using a screwdriver, remove the screws holding the fan grille to the top panel *(above)*, and lift it off. This grille provides access to the fan, fan motor and compressor.

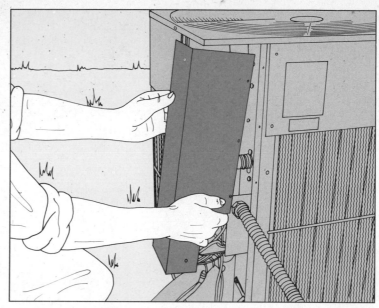

Removing the control box cover. Turn off power to the condenser unit at the main service panel *(page 134)* and unit disconnect switch. The control box is located near the entry point of the power cable and refrigerant lines. Using a screwdriver, remove the screws from the edges of the cover. For the control box shown here, pull the cover out, then down, to take it off *(above)*. The control box contains the capacitor, contactor, high-pressure switch and wiring connections. The condenser-unit wiring diagram is on the back of the control box cover.

MAINTAINING THE CONDENSER UNIT

Leveling the condenser unit. For proper operation and longer life, the condenser unit should rest on a level slab. Each spring, check whether the unit has shifted: Rest a carpenter's level across the top of the unit *(above)*, first one direction, then the other. The bubble should be centered in the level each time; if not, the mounting slab has settled. With a helper, use a pry bar to lift the low edge of the slab and prop it with gravel or sand, or have a professional adjust it with concrete. Test its level again.

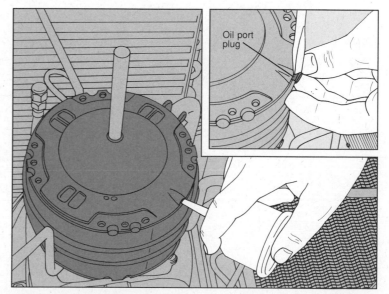

Oil port plug

Oiling the fan motor. Oil the motor once before each cooling season. Turn off power to the condenser unit at the main service panel *(page 134)* and outdoor-unit disconnect switch. Remove the top grille *(step above)* and the fan blades *(page 109)*. Locate the oil ports, small holes in the motor housing. The ports may be sealed by small plastic plugs; pry out the plugs with a screwdriver *(inset)*. Insert two or three drops of non-detergent, light machine oil into each port *(above)*; do not over-oil. Replace the plugs.

MAINTAINING THE CONDENSER COILS

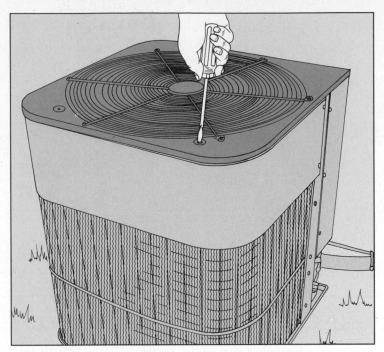

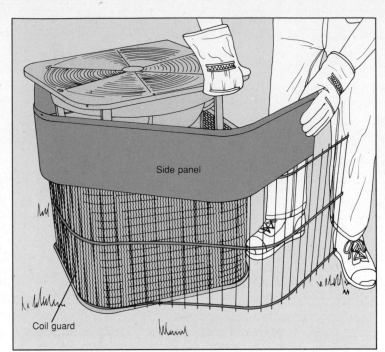

Side panel

Coil guard

1 Loosening the top panel. For minor service to the condenser coils, such as picking off leaves or straightening a few fins, work without removing the coil guard. For heavier maintenance, you may need to unscrew the top panel to release the coil guard. Turn off power to the condenser unit at the main service panel *(page 134)* and outdoor-unit disconnect switch. On the model shown, remove the screws on the top surface, near the corners *(above)*, to separate the top panel from the support rods inside.

2 Removing the coil guard. Take out the screws connecting the side panel to the top panel and to the condenser unit frame. Wearing work gloves, pull the top panel up slightly to disengage one end of the side panel and coil guard, and unwrap the panel and guard from the condenser unit frame. Your model may differ from the one shown; study the unit before disassembling it so that you do not remove more screws than necessary.

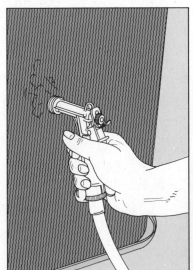

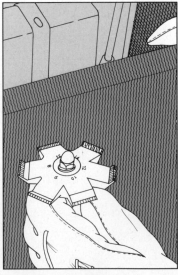

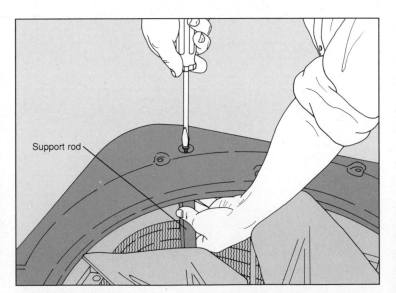

Support rod

3 Cleaning and straightening the condenser coil fins. The fan sucks in air — and debris — through the coils. Clean the coil fins before each cooling season; more often if they become clogged rapidly. Using a garden hose, spray water through the fins to dislodge dirt *(above, left)*; first spray from inside the unit outward, then spray the outside of the fins, as shown. Greasy or exceptionally dirty coils should be steam-cleaned professionally.

Check the coils for bent fins. Wearing work gloves, fit the teeth of a fin comb *(page 132)* into the fins near a damaged area. Gently pull the comb through the fins to unbend them *(above, right)*.

4 Reinstalling the panels. Fasten the screws of the side panel at one end, wrap the panel and coil guard around the unit, and then fasten the screws at the other end. Settle the top panel onto the unit and unscrew the grille *(page 107)*. Reach into the unit and hold each support rod in position while screwing the top panel to it *(above)*. Reinstall the grille.

SERVICING THE FAN BLADES

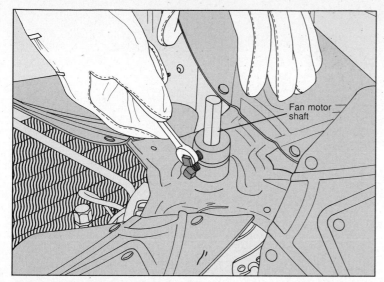

Fan motor shaft

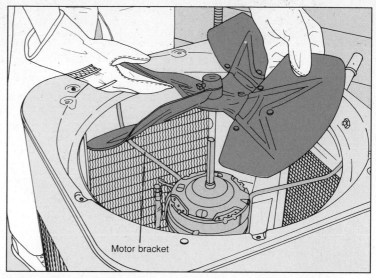

Motor bracket

1 **Inspecting the fan blades.** Loose or unbalanced fan blades can ruin the fan motor. Turn off power to the condenser unit at the main service panel *(page 134)* and outdoor-unit disconnect switch. Remove the grille *(page 107)*. If a blade is bent, do not try to straighten it; the fan blade assembly must be replaced. Wearing work gloves, spin the blades by hand. If they feel loose, or wobble as they turn, try tightening the bolts that secure the fan blades to the motor shaft *(above)*, using a wrench. If the problem persists, replace the fan blades *(step 2)*.

2 **Removing and replacing the blades.** Loosen—but do not remove—the fan blade bolts, and lift the blade assembly straight up off the motor shaft *(above)*. Purchase an identical fan blade assembly from an air conditioning supplies dealer or from the manufacturer. Settle the fan blade assembly onto the motor shaft, aligning the bolts with the flat side of the shaft. Then lift the assembly about an inch above the motor, so that the blades clear the motor bracket and their axis does not rub on the motor housing. Tighten the bolts with a wrench. Check whether the blades are loose, wobbly or striking the bracket, and adjust them.

DISCHARGING THE CAPACITOR

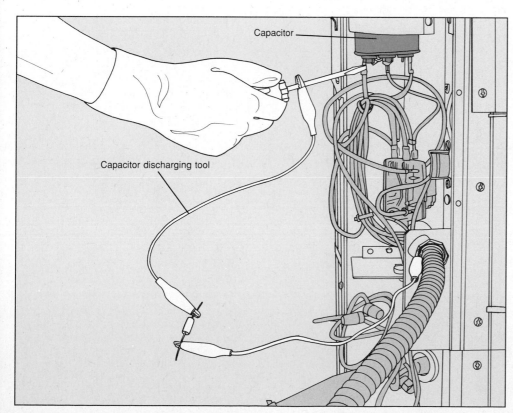

Capacitor

Capacitor discharging tool

Using a capacitor discharging tool. Caution: Capacitors may store potentially dangerous voltage. Before servicing control box components, the fan motor or the compressor, make a capacitor discharging tool *(page 140)* and discharge the capacitor. Turn off power to the condenser unit at the main service panel *(page 134)* and outdoor-unit disconnect switch. Remove the control box cover *(page 107)* to access the capacitor. Clip the free end of the capacitor discharging tool to an unpainted metal part of the unit chassis. Hold the insulated screwdriver handle in one hand and touch the blade to each capacitor terminal, one at a time, for one second *(left)*. Look for other capacitors in the unit and discharge them the same way.

TESTING AND REPLACING A CAPACITOR

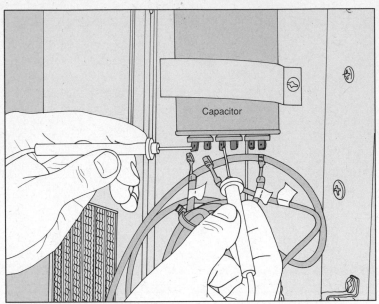

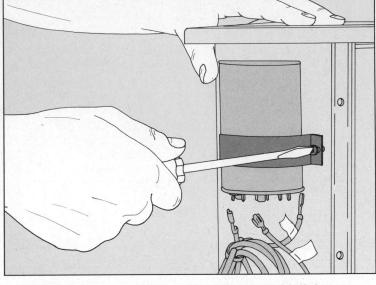

Testing and replacing the capacitor. Turn off power to the condenser unit at the main service panel *(page 134)* and outdoor-unit disconnect switch. Remove the control box cover *(page 107)* and discharge the capacitor *(page 109)*. To test a three-terminal capacitor, as shown, label and disconnect its wires. Set a multitester to RX1K to measure resistance *(page 137)*. Touch one probe to the common terminal, marked "C", and touch the other probe to one of the other terminals *(above, left)*. The multitester needle should swing toward zero resistance, then slowly move across the scale toward infinity. Test between "C" and the third terminal; if the

needle responds the same way, the capacitor is good. If, for either test, the needle swings to zero and stays there, or it does not move at all, replace the capacitor. If the condenser unit has two, two-terminal capacitors, discharge them *(page 109)* and test each as on page 95. If a capacitor is faulty, unscrew the capacitor bracket *(above, right)* and slide out the capacitor. Purchase an identical replacement from a heating and cooling supplies dealer. Install the new capacitor in the bracket and screw the bracket in place. Reconnect the wires to the correct terminals.

SERVICING THE CONTACTOR

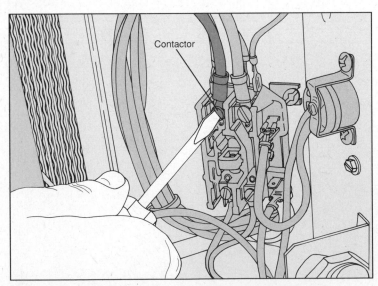

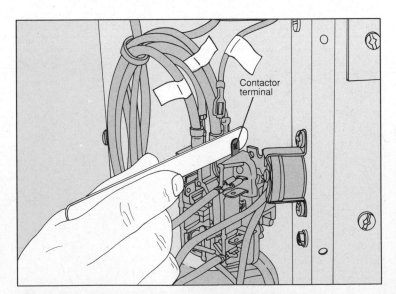

1 **Disconnecting the wires.** Turn off power to the condenser unit at the main service panel *(page 134)* and outdoor-unit disconnect switch. Remove the control box cover *(page 107)* and discharge the capacitor *(page 109)*. Label the contactor wires *(page 135)*. Disconnect the wires from the contactor, unscrewing screw terminals *(above)*, and pulling off spade connectors with long-nose pliers. For easier access, unscrew the contactor bracket and remove the contactor from the control box.

2 **Cleaning the terminals.** A contactor that doesn't work properly may simply be dirty. Inspect the contactor terminals for dirt or corrosion, and rub away any deposits with an ice cream stick *(above)*. Check the wire connectors you removed in step 1, and replace any that are loose-fitting or damaged *(page 139)*. Reconnect the wires.

SERVICING THE CONTACTOR (continued)

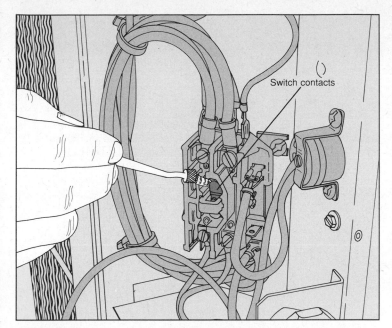

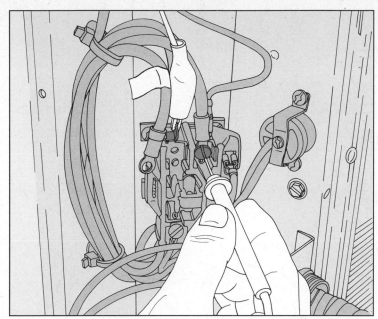

3 **Cleaning the switch contacts.** Inspect the reed-style switch contacts. If the contacts are stuck together or the spring is missing, replace the contactor *(step 6)*. Clean the contacts with contact cleaner solution, available at an electronics supply store. Use a small tool such as a dental brush *(above)* or a foam swab to reach the area where the contacts meet. Reconnect any loose wires, reassemble the unit and turn on the power; if the contactor still doesn't work, test it *(step 4)*.

4 **Testing the contactor with the switch contacts open.** Turn off power to the unit and gain access to the contactor *(step 1)*. Disconnect the wires from one upper screw terminal. Set a multitester to test continuity *(page 136)*. Clip one probe to the disconnected terminal, and touch the other probe to the screw terminal next to it *(above)*. There should be no continuity. Test the lower pair of terminals the same way, first disconnecting the wires from one terminal. If either test shows continuity, the contactor is faulty; replace it *(step 6)*. If it is OK, test with the contacts closed *(step 5)*.

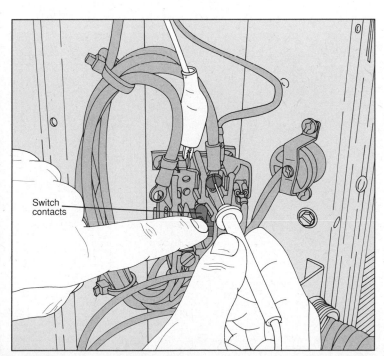

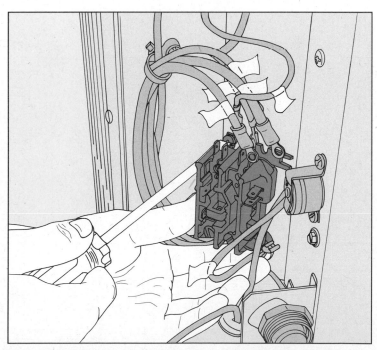

5 **Testing the contactor with the switch contacts closed.** With power to the unit still off, touch a multitester probe to each upper terminal as in step 4; but this time, depress the switch to close the contacts *(above)*. There should be continuity. Test the lower pair of terminals the same way. If either test does not show continuity, replace the contactor *(step 6)*. If the contactor is OK, reconnect the wires, reassemble the unit and turn on the power.

6 **Replacing the contactor.** Label and disconnect all contactor wires *(page 135)*. Remove the mounting screws holding the contactor bracket to the control box wall *(above)* and take out the contactor. Purchase an identical replacement contactor from a heating and cooling supplies dealer or the manufacturer. Screw the new contactor in place and reconnect the wires. Reassemble the unit and turn on the power.

SERVICING THE OUTDOOR TEMPERATURE SENSOR

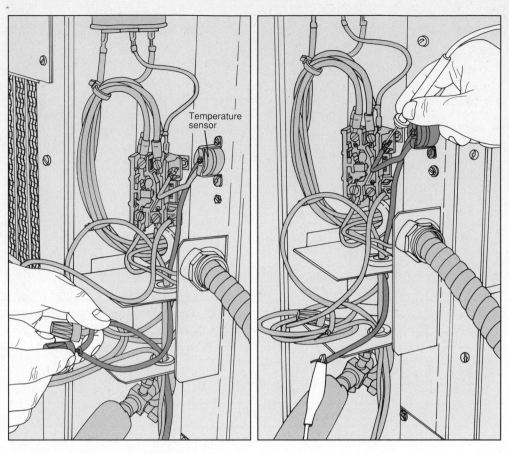

Temperature sensor

1 **Testing the outdoor temperature sensor.** Turn off power to the condenser unit at the main service panel *(page 134)* and outdoor-unit disconnect switch. Remove the control box cover *(page 107)* and discharge the capacitor *(page 109)*. Label all sensor wires for correct reconnection *(page 135)*. Disconnect a wire by unscrewing the wire cap *(far left)* or pulling off the spade connector. Set a multitester to test continuity *(page 136)*. The sensor must be tested at temperatures above 45°F and below 45°F. First, clip a multitester probe to one sensor lead and touch the other probe to the other lead *(near left)*. There should be continuity if the outdoor temperature is above 45°F; if not, replace the sensor *(step 2)*. If the sensor passes this test, next remove it from the control box *(step 2)*, put it in the freezer for 15 minutes and immediately test again. This time, there should be no continuity. If the sensor fails either test, replace it *(step 2)*.

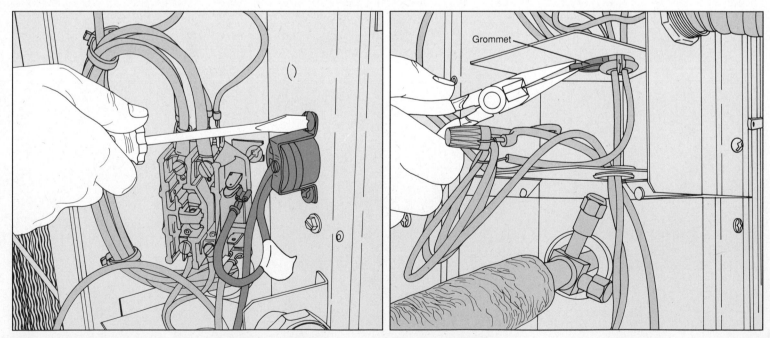

Grommet

2 **Removing and replacing the sensor.** With the sensor wires labeled and disconnected *(step 1)*, unscrew the bracket holding the sensor to the control box wall *(above, left)*. If a grommet secures the wiring, unfasten it; you may need to snap off a metal clip using long-nose pliers *(above, right)*. Purchase an identical replacement sensor from a heating and cooling supplies dealer. Screw the sensor bracket in place and secure the wires in the grommet. Reconnect the wires, reassemble the unit and turn on the power.

SERVICING THE HIGH-PRESSURE SWITCH

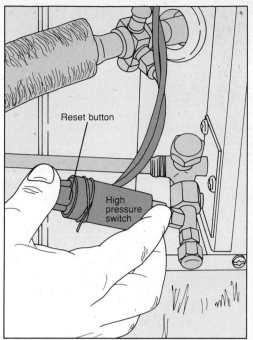

Reset button

High pressure switch

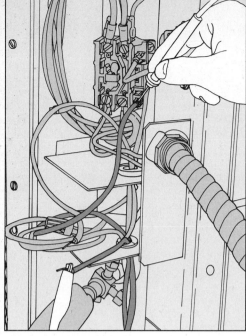

Resetting and testing the high-pressure switch. Turn off power to the condenser unit at the main service panel *(page 134)* and outdoor-unit disconnect switch. Remove the control box cover *(page 107)* and discharge the capacitor *(page 109)*. Allow the unit to cool several hours. The high-pressure switch cuts off power to the condenser unit if pressure in the refrigerant line becomes too high. This may be a momentary problem; restart the unit by depressing the reset button *(far left)*, then restoring the power. If the high-pressure switch continually turns off the unit, the switch may be faulty. To test the switch, let the unit cool several hours, then reset the switch. Label its wires *(page 135)* and disconnect one of them. Set a multitester to test continuity *(page 136)*, and touch a probe to each wire end *(near left)*. There should be continuity. If the switch tests OK, the problem is in the refrigeration system; have it serviced professionally. If the switch tests faulty, have it replaced professionally.

TESTING AND REPLACING THE FAN MOTOR

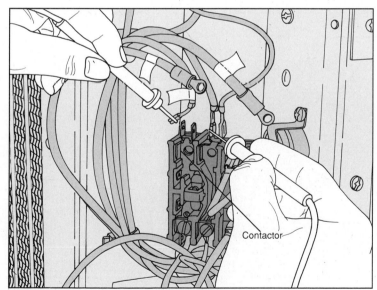

Contactor

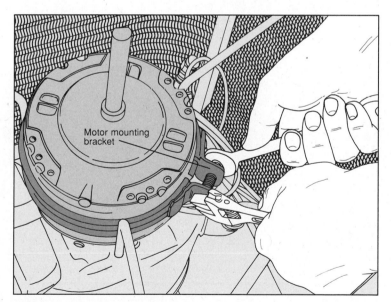

Motor mounting bracket

1 **Testing the fan motor.** Turn off power to the condenser unit at the main service panel *(page 134)* and outdoor-unit disconnect switch *(page 135)*. Take off the fan grille and control box cover *(page 107)* and the fan blades *(page 109)*. Discharge the capacitor *(page 109)*. Trace the wires leading from the motor to the contactor; on a one-speed motor, shown here, there should be two wires. Disconnect one of the fan wires from the contactor; disconnect other wires in the way, if necessary. Set a multitester to test continuity *(page 136)* and touch a probe to each fan wire *(above)*. The tester should show continuity. Then touch one probe to the motor housing and the other probe to each fan wire, in turn. There should be no continuity. If the fan motor fails any test, replace it *(step 2)*.

2 **Loosening the motor bracket.** In the model shown here, the fan motor is suspended in a bracket over the center of the condenser. Before unbolting the bracket, pull the motor wires through from the control box; you may have to release some wire-retaining bands first. Loosen the mounting bracket bolt using two wrenches, one wrench to hold the head of the bolt in place and the other to turn the nut *(above)*. Do not remove the nut.

TESTING AND REPLACING THE FAN MOTOR (continued)

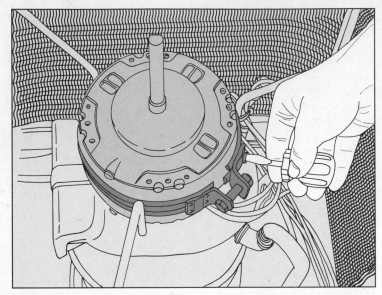

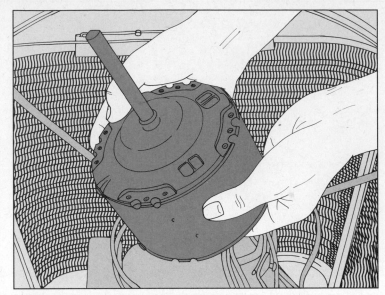

3 **Loosening the retaining screw.** There may be a screw in the mounting bracket that is set against the motor to hold it steady. Loosen the screw with a short screwdriver *(above)*.

4 **Removing the motor.** With the motor loosened from its bracket and its wires freed from the control panel, lift out the motor. The motor is heavy; hold it firmly by the housing *(above)*, not the shaft. Purchase an exact replacement motor from the manufacturer or a heating and cooling supplies dealer; take the old motor with you when ordering. Install the new motor, reversing the steps taken to remove it. Reinstall the fan blades *(page 109)*, reassemble the unit and turn on the power.

SERVICING THE COMPRESSOR

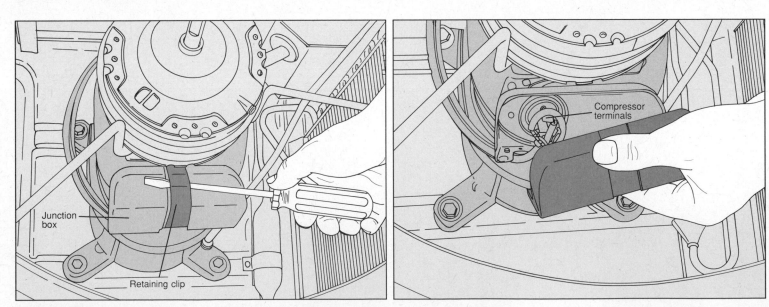

Junction box

Retaining clip

Compressor terminals

1 **Accessing the junction box.** Turn off power to the condenser unit at the main service panel *(page 134)* and the outdoor-unit disconnect switch. Remove the grille and control box cover *(page 107)*. Take out the fan blades and discharge the capacitor *(page 109)*. The junction box, on the side of the compressor, houses the compressor motor terminals. The junction box cover is held in place by a retaining clip. Wedge a screwdriver beween the clip and the cover *(above, left)*, and twist it to snap off the clip. Lift off the junction box cover *(above, right)*. The device with three wires connected to it is the set of compressor motor terminals. Some compressors have an overload protector, also housed in the junction box. Service it as described on page 128.

SERVICING THE COMPRESSOR (continued)

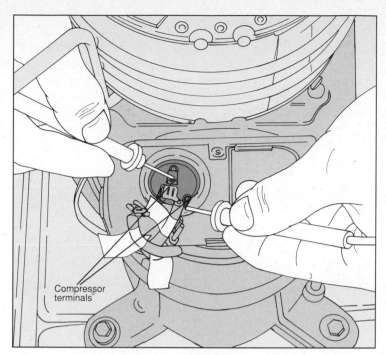

Compressor
terminals

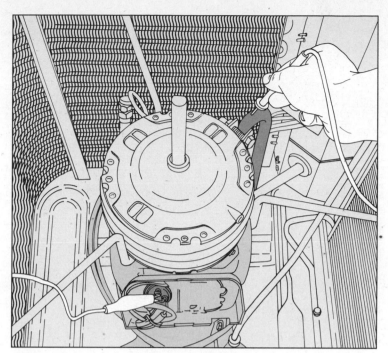

2 **Testing the compressor motor.** Label and disconnect the wires *(page 135)*. Set a multitester to test continuity *(page 136)*. Touch one probe to one compressor motor terminal, and touch the other probe to another terminal *(above)*. There should be continuity. Repeat the test for all three possible combinations of terminals. If there is continuity, the compressor motor windings are working; go to step 3. If there is no continuity in any one of the tests, the compressor is faulty; call for service.

3 **Testing the compressor motor for ground.** Allow the compressor to cool for three to four hours. Set a multitester to test continuity *(page 136)*. Touch one probe to a compressor motor terminal and touch the other probe to a clean, unpainted metal surface of the compressor housing, or to a copper refrigerant line leading from the compressor *(above)*. There should be no continuity. Test all three terminals this way. If all tests show no continuity, the compressor is OK; reconnect the wires and go to step 4. If any test shows continuity, the motor is grounded and the compressor must be replaced; call for service.

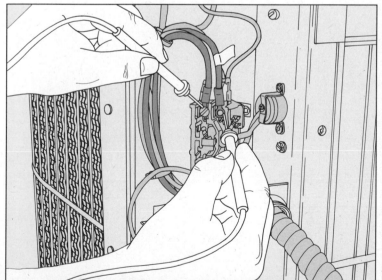

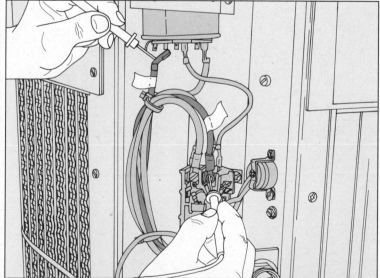

4 **Testing the compressor wiring.** A minor problem with its electrical wiring or connections can make a compressor malfunction. After determining that the compressor is OK, test the compressor wires. Trace the three compressor wires to the control box; two go to the contactor and one to the capacitor. Label and disconnect the wires *(page 135)*. Set a multitester to test continuity *(page 136)*. Touch a probe to each compressor wire that leads to

the contactor *(above, left)*; there should be continuity. Next, touch one probe to the compressor wire that leads to the capacitor *(above, right)*, and touch the other probe to each of the other two wires, in turn. In both cases, there should be continuity. Replace a faulty wire with one of identical rating and install the correct terminal connectors *(page 138)*. Reassemble the condenser unit.

WINDOW AIR CONDITIONERS

In addition to cooling a room, a window air conditioner filters, dehumidifies and circulates room air. A unit runs on either 120- or 240-volt electrical current, depending on its size and rating. Current enters the unit through a three-prong power cord and passes to the selector switch, then through electricity-storing capacitors to the motor and compressor.

Flowing through the air conditioner is a pressurized refrigerant, alternately in gas and liquid form. When the air conditioner is switched on, the compressor sucks in refrigerant gas and pressurizes it, raising its temperature. The heated, high-pressure gas then travels to the condenser coils outdoors, where the fins give off its heat to the surrounding cooler air as it condenses into a liquid. The liquid refrigerant next travels indoors to the evaporator coils where, under reduced pressure, it vaporizes into a gas, absorbing heat from the air in the room. The blower pulls room air through the air filter and across the evaporator coils, where it is cooled, then blown back into the room through louvers in the front panel. Water vapor from the cooled air condenses on the coils, drips into a drain pan and is splashed by the slinger ring to cool the condenser coils.

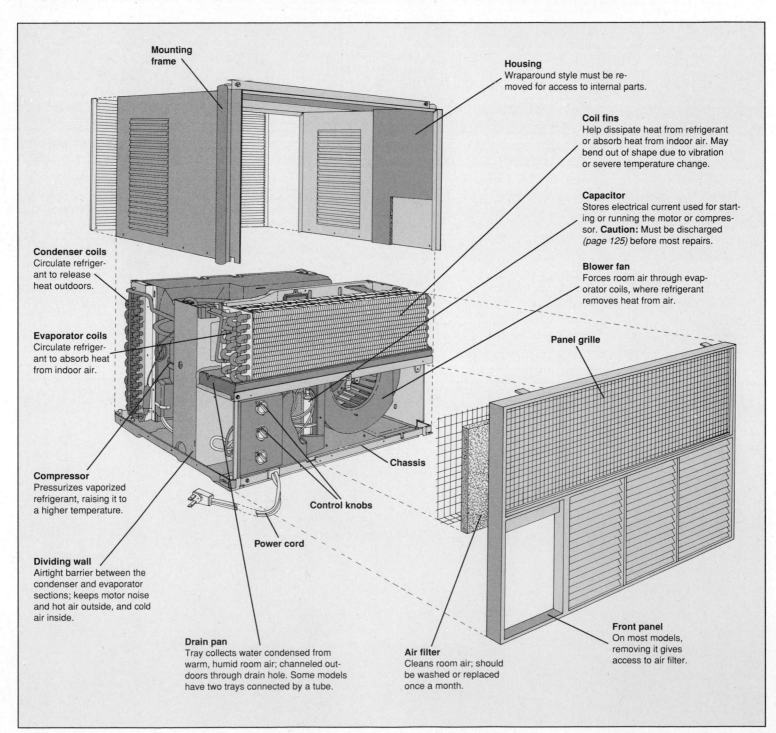

Mounting frame

Housing
Wraparound style must be removed for access to internal parts.

Coil fins
Help dissipate heat from refrigerant or absorb heat from indoor air. May bend out of shape due to vibration or severe temperature change.

Capacitor
Stores electrical current used for starting or running the motor or compressor. **Caution:** Must be discharged *(page 125)* before most repairs.

Blower fan
Forces room air through evaporator coils, where refrigerant removes heat from air.

Condenser coils
Circulate refrigerant to release heat outdoors.

Evaporator coils
Circulate refrigerant to absorb heat from indoor air.

Panel grille

Chassis

Compressor
Pressurizes vaporized refrigerant, raising it to a higher temperature.

Control knobs

Power cord

Dividing wall
Airtight barrier between the condenser and evaporator sections; keeps motor noise and hot air outside, and cold air inside.

Drain pan
Tray collects water condensed from warm, humid room air; channeled outdoors through drain hole. Some models have two trays connected by a tube.

Air filter
Cleans room air; should be washed or replaced once a month.

Front panel
On most models, removing it gives access to air filter.

With proper maintenance, the average window air conditioner will serve you well for ten years. Once a year, clean the grille *(page 121)*, clean the coils and straighten the fins *(page 121)*, clean the drain system *(page 123)* and lubricate the motor *(page 124)*. Most importantly, clean or replace your air filter *(page 120)* at least once a month.

Apart from changing the air filter, all air conditioner repairs require first removing the unit from the window; since it is quite heavy, have someone help you. Some models also require taking off the housing *(page 120)*. Models vary in other ways, such as the number of capacitors they use, the settings available on the selector switch, and the location of the air filter, overload protector or other components.

Follow all safety instructions in this chapter and in the Emergency Guide *(page 8)*. Before beginning any repair or inspection, unplug the power cord. **Caution:** Wherever recommended, make a capacitor discharger *(page 140)* and discharge all capacitors *(page 125)*. Avoid handling the fins, coils or refrigerant lines; these carry high-pressure refrigerant, and should be serviced only by a professional.

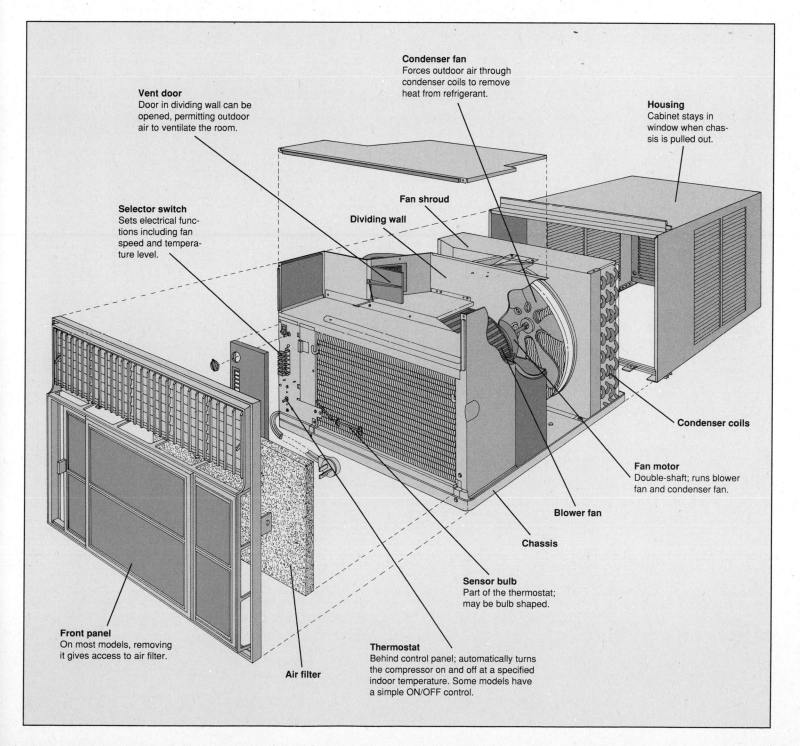

Condenser fan
Forces outdoor air through condenser coils to remove heat from refrigerant.

Vent door
Door in dividing wall can be opened, permitting outdoor air to ventilate the room.

Housing
Cabinet stays in window when chassis is pulled out.

Fan shroud

Dividing wall

Selector switch
Sets electrical functions including fan speed and temperature level.

Condenser coils

Fan motor
Double-shaft; runs blower fan and condenser fan.

Blower fan

Chassis

Sensor bulb
Part of the thermostat; may be bulb shaped.

Front panel
On most models, removing it gives access to air filter.

Air filter

Thermostat
Behind control panel; automatically turns the compressor on and off at a specified indoor temperature. Some models have a simple ON/OFF control.

TROUBLESHOOTING GUIDE

SYMPTOM	POSSIBLE CAUSE	PROCEDURE
Air conditioner does not run at all	No power to unit	Replace fuse(s) or reset circuit breaker (p. 134) □○
	Overload protector tripped	Wait for air conditioner to cool, and turn it on again; test overload protector (p. 128) ◓●
	Filter clogged	Wash or replace filter (p. 120) □○
	Selector switch faulty	Test selector switch (p. 126) ◓●▲ ; replace if necessary
	Power cord faulty	Test power cord (p. 124) ◓●▲ ; replace if necessary
Air conditioner repeatedly blows fuse(s) or trips circuit breaker	Capacitor faulty	Discharge and test capacitor (p. 125) ◓●▲ ; replace if necessary
	Compressor faulty	Test compressor (p. 127) ◓●▲ ; call for service
	Fan motor faulty	Test fan motor (p. 129) ◓○▲ ; replace (p. 129) ■● if necessary
	Coils dirty or fins flattened	Clean coils and straighten fins (p. 120) □○▲
Fan runs but air conditioner does not cool at all	Thermostat set above room temperature	Set air conditioner thermostat below room temperature
	Compressor faulty	Test compressor (p. 127) ◓●▲ ; call for service if necessary
	Thermostat faulty	Test thermostat (p. 126) □●▲ ; replace if necessary
	Selector switch faulty	Test selector switch (p. 126) ◓●▲ ; replace if necessary
	Capacitor faulty	Discharge and test capacitor (p. 125) ◓●▲ ; replace if necessary
	Overload protector faulty	Test overload protector (p. 128) ◓●▲ ; replace if necessary
Evaporator coils frost up but no air circulates	Selector switch faulty	Test selector switch (p. 126) ◓●▲ ; replace if necessary
	Capacitor faulty	Test capacitor (p. 125) ◓●▲ ; replace if necessary
	Fan motor bearings dry	Lubricate motor (p. 124) ◓●
	Fan motor faulty	Test fan motor (p. 129) ◓○▲ ; replace (p. 129) ■● if necessary
Air conditioner short cycles (turns on and off repeatedly)	Capacitor faulty	Discharge and test capacitor (p. 125) ◓●▲ ; replace if necessary
	Thermostat sensor bulb touching coil	Reposition sensor bulb (p. 122) □○
	Thermostat faulty	Test thermostat (p. 126) □●▲ ; replace if necessary
	Air flow through condenser obstructed	Remove obstruction; clean coils and straighten fins (p. 121) □○▲
Air conditioner does not cool sufficiently	Thermostat set too high	Set air conditioner thermostat lower
	Selector switch set to ventilate	Set switch to circulate; seal leaks where housing meets window
	Cool air deflected back through front panel	Move any obstructions such as draperies or furniture
	Grille on front panel dirty	Clean grille (p. 120) □○
	Air filter dirty	Clean or replace filter (p. 120) □○
	Refrigerant tubing blocked inside	Check coils for heavy icing; call for service
	Refrigerant leaking	Check for oily leaks around air conditioner; call for service
	Low fan speed allowing evaporator coils to ice up	Set fan to higher speed
	Loose blower obstructing air flow	Tighten blower setscrew (p. 129) ◓○
Air conditioner is noisy or vibrates	Front panel or housing loose	Tighten loose screws or bolts
	Thermostat sensor bulb touching coil	Reposition sensor bulb (p. 122) □○
	Compressor junction-box cover loose	Replace the junction-box clip (p. 127) ◓○
	Blower or condenser fan loose	Tighten setscrew on blower or fan (p. 130) ◓○
	Fan motor bearings dry	Lubricate motor (p. 124) ◓○
	Window installation loose	Tighten mounting screws; secure sliding side panels
Air conditioner drips water indoors	Air conditioner improperly positioned	Tilt air conditioner down slightly toward the outside
	Drain system blocked	Clear drain system (p. 123) ◓●
Condensation forms on front panel	Extreme humidity	Wipe with cloth or allow to evaporate; install room humidifier
Oily leak inside or outside	Refrigerant leaking	Call for service

DEGREE OF DIFFICULTY: □ Easy ◓ Moderate ■ Complex
ESTIMATED TIME: ○ Less than 1 hour ◓ 1 to 3 hours ● Over 3 hours   ▲ Special tool required

ACCESS TO THE INTERNAL COMPONENTS

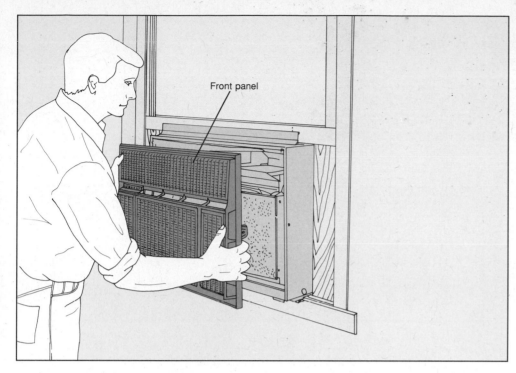

Front panel

1 Removing the front panel. Unplug the air conditioner. If you are just changing the air filter, leave the air conditioner in the window, as shown; for most other repairs, remove the air conditioner from the window *(step 2)*. To take off the front panel, remove any bolts or screws, then pull the front panel straight off. If the panel is secured by clips, simply grip its sides and snap it off *(left)*.

Chassis

2 Taking the air conditioner out of the window. Most larger, heavier air conditioners, such as the one shown, have a slide-out chassis that fits into a cabinet bolted permanently to the window frame. Many small air conditioners have sliding panels at the sides, and may or may not be supported by a bracket outside the window. To remove a small air conditioner, have a helper support it while you carefully open the window. **Caution:** The air conditioner is very heavy and can easily tip backward. Collapse the side panels.

Place a sturdy table next to the window, under the air conditioner, and work with a helper to remove it. Keeping your back straight and your knees bent to avoid muscle strain, slide the chassis—or the entire unit—out of the window onto the table. If the air conditioner has a wraparound housing, remove it *(step 3)*.

ACCESS TO THE INTERNAL COMPONENTS (continued)

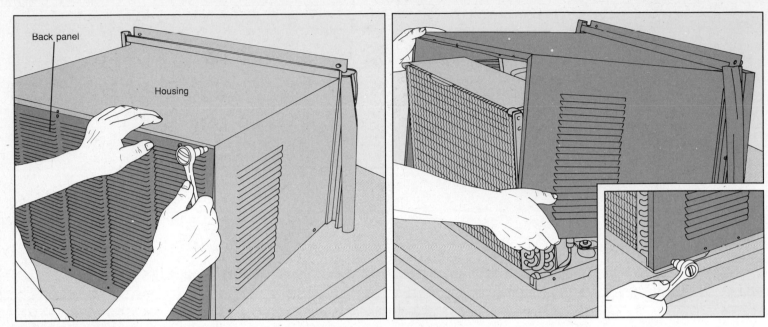

3 **Removing a wraparound housing.** Most small air conditioners have a back panel secured to the chassis with bolts or sheet metal screws. Use a socket wrench to remove bolts *(above, left)* or a screwdriver to remove screws; if they are hard to remove, apply several drops of penetrating oil to free them. Next, remove the bolts or screws from the top and sides of the housing *(inset)* and, on some models, at the front. Grasp the bottom edges of the housing and pull it up off the air conditioner *(above, right)*. Set the housing aside, away from the work area.

MAINTAINING THE AIR FILTER AND GRILLE

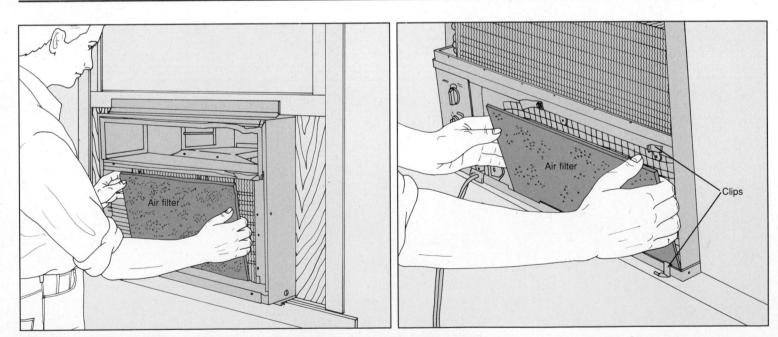

1 **Removing the air filter.** Unplug the air conditioner and remove the front panel *(page 119)*. On some models, the filter is located either in front of the evaporator coils or on the back of the front panel; unfasten the retaining clips holding the filter in place and remove the filter *(above, left)*. On other models, the filter is mounted in front of the blower and below the evaporator coils; unfasten the filter from its clips and pull it out *(above, right)*.

MAINTAINING THE AIR FILTER AND GRILLE (continued)

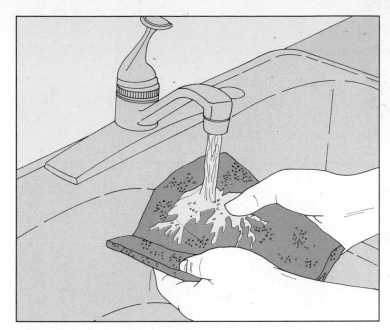

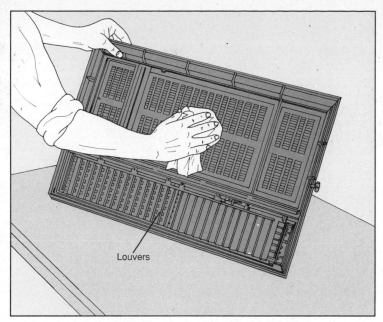

Louvers

2 **Washing or replacing the filter.** Each month during the cooling season, wash or replace the filter. If the filter is not washable or it is torn, replace it with an identical filter. If the filter is washable, vacuum off surface dirt, then wash the filter in a detergent-and-water solution, rinse it with fresh water *(above)* and wring it dry.

3 **Cleaning and replacing the front panel.** Use a moist cloth or stiff-bristled brush to wipe accumulated dust off both sides of the grille and louvers of the front panel *(above)*. Wash the panel in a detergent-and-water solution to remove greasy dirt, rinse it in clear water and dry it. Reinstall the front panel and plug in the air conditioner.

MAINTAINING THE COILS

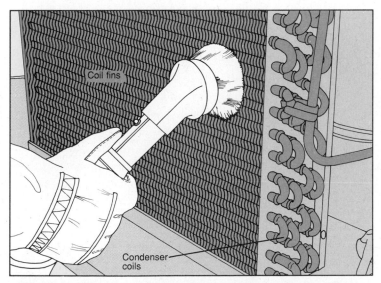

Coil fins

Condenser coils

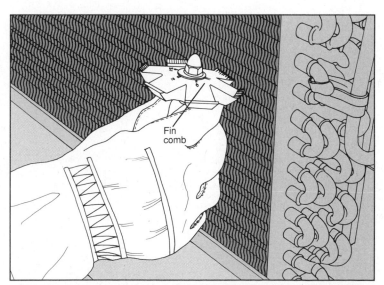

Fin comb

1 **Vacuuming the coil fins.** Unplug the air conditioner, remove it from the window and take off the wraparound housing, if any *(page 119)*. Wearing work gloves to protect your hands, use a vacuum cleaner with an upholstery-brush attachment to vacuum dirty condenser coil fins *(above)* and evaporator coil fins. If the coils are greasy or exceptionally dirty, have them steam-cleaned by a professional.

2 **Straightening the coil fins.** With the air conditioner unplugged, use a multi-headed fin comb *(page 132)* to straighten bent coil fins. Determine which head of the comb corresponds to the spacing of fins on the coils; the teeth on the head should fit easily between the fins. Wearing heavy work gloves, gently fit the fin comb teeth between the coil fins, into an undamaged section above the area to be straightened. Pull the fin comb down, sliding it through the damaged area *(above)*; at the same time, comb out any lodged debris. **Caution:** Do not use a knife or screwdriver to straighten fins; they can damage the coils.

REMOVING THE CONTROL PANEL

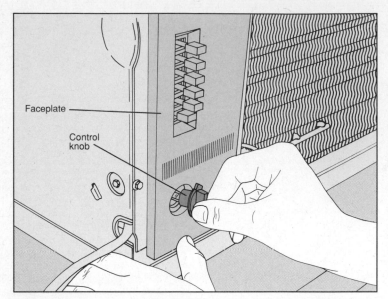

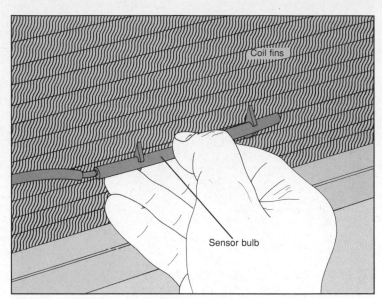

1 **Removing the control knobs.** Unplug the air conditioner and remove it from the window, then take off the wraparound housing *(page 119)*. Pull off any control knobs too big to fit through the faceplate *(above)*. If the control knobs cannot be removed easily, use a screwdriver to pry them away from the faceplate; if the knobs are secured by setscrews, use a screwdriver or hex wrench to remove the screws. Pull the faceplate off the front of the control panel.

2 **Removing the thermometer sensor bulb.** Take out the air filter *(page 120)*. Locate the sensor bulb, which may be a copper tube as shown above, or may be bulb-shaped. Gently pull the bulb off its bracket *(above)*. If the sensor bulb was touching the evaporator coils, adjust it and reassemble the air conditioner.

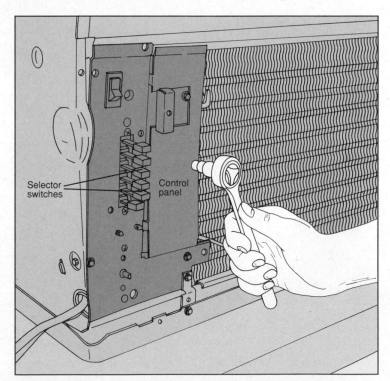

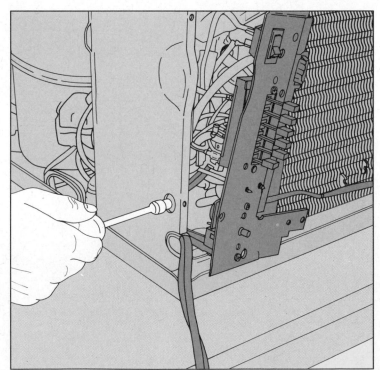

3 **Unfastening the control panel.** Remove the control panel's screws with a screwdriver, or bolts with a socket wrench *(above)*. Pull the control panel out slightly to access the wiring connected to the selector switch, power cord, thermostat and, on some models, the vent door control.

4 **Releasing the power cord.** The power cord is secured to the inside of the cabinet by a bracket or grommet. Free the control panel by unscrewing the bracket screw *(above)* or prying out the grommet. Pull the control panel away from the air conditioner, taking care not to damage or pull off the wires.

SERVICING THE DRAIN SYSTEM

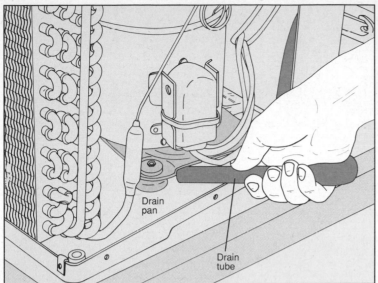

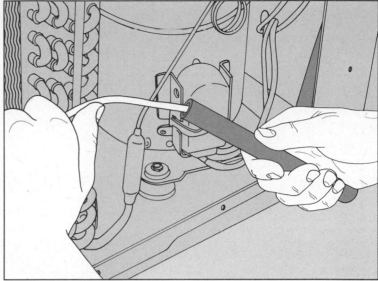

1 **Clearing the drain tube.** Unplug the air conditioner and access the internal components *(page 119)*. If your model has an evaporator drain pan and a condenser drain pan, locate the rubber or plastic drain tube connecting them. Pull the tube out from under the compressor base *(above, left)* and run a heavy wire through the tube to dislodge obstructions *(above, right)*. Flush the tube with one tablespoon chlorine bleach in one-half cup water to prevent algae formation. If the air conditioner does not have a drain tube, use a cloth to wipe clean the drain channels molded into the drain pan.

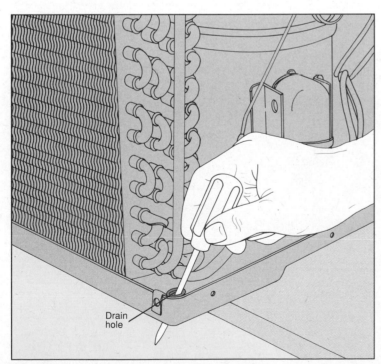

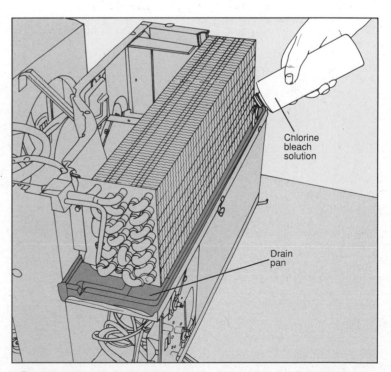

2 **Clearing the drain hole.** Locate the drain hole leading out of the chassis. If the hole is blocked, use a screwdriver or heavy wire to clear the hole *(above)*.

3 **Cleaning the drain pan.** Place a bucket under the drain hole to catch overflow. Prevent algae formation in the drain pans by flushing each pan with a solution of one cup chlorine bleach and one cup water, then rinsing it with clear water.

LUBRICATING THE FAN MOTOR

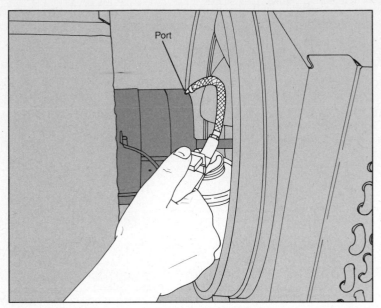

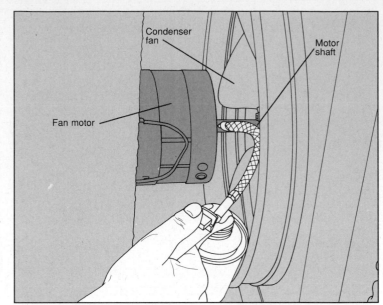

1 **Oiling the motor.** Unplug the air conditioner and access the internal components *(page 119)*. Look for oil ports on the motor housing; if there are none, go to step 2. If the motor has ports, use a screwdriver to pry off the oil port plugs. Insert three drops of SAE 10 non-detergent oil into each port. **Caution:** Excess oil can damage the motor. If the ports are difficult to reach, use an oil can with a long, flexible spout *(above)*.

2 **Lubricating the motor shaft.** Turn the motor shaft manually; if it squeaks, insert two drops of SAE 10 non-detergent oil along the shaft between the motor housing and the condenser fan *(above)*. Turn the condenser fan back and forth to work the oil into the motor. Reassemble and reinstall the air conditioner.

SERVICING THE POWER CORD

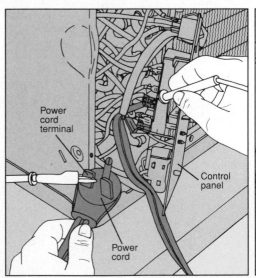

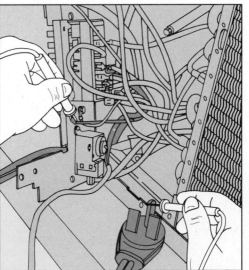

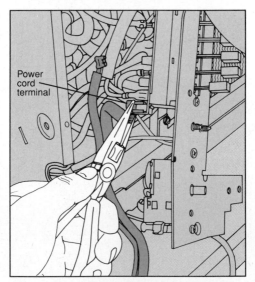

1 **Testing the power cord.** To access the internal power-cord connections, remove the control panel *(page 122)*. If the plug or cord are discolored or damaged, replace the cord *(step 2)*. Set a multitester to the RX1K scale to measure continuity. Clip one probe to one plug prong, and touch the other probe to one power cord terminal, then the other *(above, left)*. There should be a continuity reading with only one terminal. Repeat the test for the other prong; there should be continuity with the other terminal. If the cord fails the test, replace it. If the cord tests OK, attach one tester probe to the ground prong, and the other tester probe to the ground wire at the control panel *(above, right)*. If there is continuity, the ground wire is OK; If there is no continuity, replace the cord.

2 **Replacing the power cord.** Pull the power-cord spade lugs off the selector switch terminals *(above)*, and unscrew the ground wire from the control panel. Install a new three-prong plug and cord of the same electrical rating (120 or 240 volts) and wire gauge as the old one. **Caution:** Be sure to use a compatible cord to avoid the risk of fire or shock.

DISCHARGING THE CAPACITORS

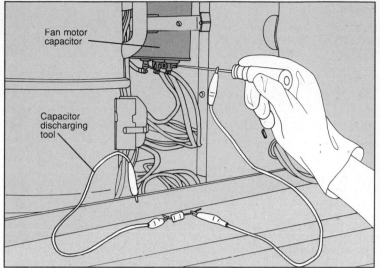

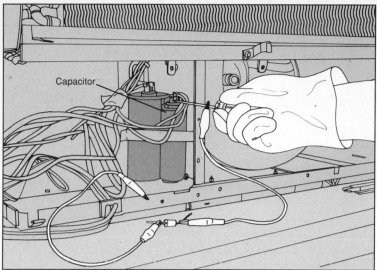

Locating and discharging capacitors. Caution: Capacitors store potentially dangerous voltage. Before servicing any internal components, or working near a capacitor, make a capacitor discharging tool *(page 140)* and discharge all the capacitors. Capacitors are mounted either behind the dividing wall near the compressor and fan motor *(above, left)* or behind the control panel *(above, right)*. Unplug the air conditioner and access the internal components *(page 119)*. Remove the control panel *(page 122)* if necessary.

Identify the air conditioner's two or three capacitors: Capacitors may vary in shape and size *(page 140)*, may have two or three terminals and may or may not have a bleed resistor. Discharge a capacitor by clipping one end of the capacitor discharger to the air conditioner chassis. Hold the insulated screwdriver handle of the capacitor discharger in one hand and touch each of the terminals with the tip of the screwdriver blade for one second *(above, left and right)*. Discharge all other capacitors in the air conditioner.

SERVICING CAPACITORS

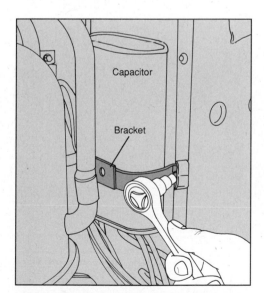

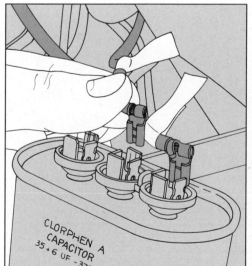

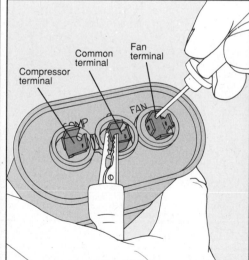

1 Removing the capacitor. With the capacitor discharged *(step above)*, use a screwdriver to remove screws or a socket wrench to remove bolts *(above)* holding the bracket in place. Being careful not to touch the terminals with your hands, pull the capacitor out of the air conditioner.

2 Disconnecting the capacitor. Disconnect the wire leads from the capacitor terminals *(above, left)*. Label the wires for correct reconnection *(page 135)*. Set a multitester to the RX1K scale to measure resistance *(page 137)*. If the capacitor has two terminals, test it as shown on page 95. To test a three-terminal capacitor, place one multitester probe on the terminal marked "C" and touch the other probe to one of the other terminals. The needle should swing toward zero resistance, then slowly move one-third to halfway across the scale toward infinity. Test between "C" and the third terminal. If the needle responds the same way, the capacitor is good; test the other capacitors. If, for either test, the needle swings to zero resistance and stays there, or it does not move at all, replace the capacitor with an identical part.

SERVICING THE THERMOSTAT

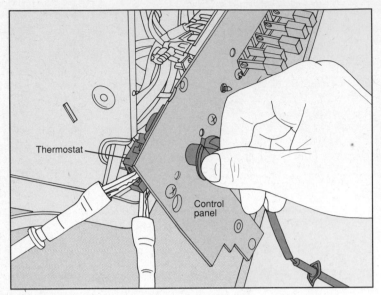

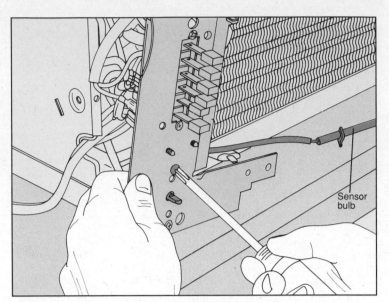

1 **Testing the thermostat.** Unplug the air conditioner, remove the control panel *(page 122)* and discharge the capacitors *(page 125)*. Locate the thermostat, which is connected to the sensor bulb; disconnect the two thermostat wire leads. Set a multi-tester *(page 136)* to the RX1K setting to measure continuity. With the thermostat on its highest setting, touch one probe to each of the terminals; there should be no continuity. Repeat the test at the lowest setting; there should be continuity. If the thermostat fails either test, replace it.

2 **Removing the thermostat.** If the thermostat has a sensor bulb, disconnect it *(page 122)*. Remove the mounting screws holding the thermostat to the control panel *(above)*. Take out the thermostat and sensing bulb, and install an exact replacement, available from an air-conditioning parts dealer or from the manufacturer. Reassemble the air conditioner and plug it in.

SERVICING THE SELECTOR SWITCH

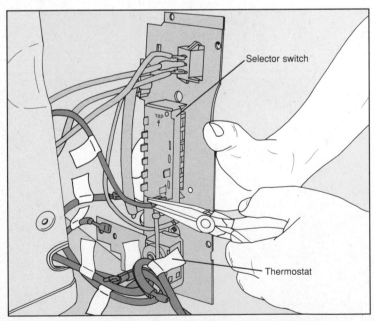

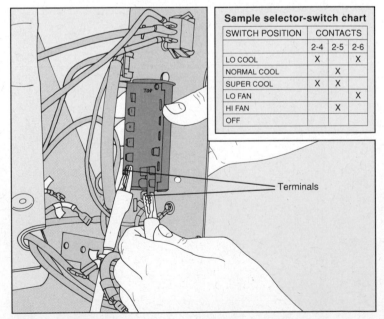

Sample selector-switch chart			
SWITCH POSITION	CONTACTS		
	2-4	2-5	2-6
LO COOL	X		X
NORMAL COOL		X	
SUPER COOL	X	X	
LO FAN			X
HI FAN		X	
OFF			

1 **Disconnecting the wire leads.** Unplug the air conditioner, remove the control panel *(page 122)* and discharge the capacitors *(page 125)*. Use long-nose pliers to detach the wire leads from the selector switch. Label *(page 135)* each wire as you remove it, and draw a diagram of the switch terminals, for correct reconnection.

2 **Identifying the terminals.** Look for the selector-switch wiring diagram, located near the control panel. The diagram includes a small chart like the sample shown above *(inset)*. It shows the correct position (open or closed) for each switch contact at various selector-switch settings. Referring to your chart, identify the pairs of terminals to be tested.

SERVICING THE SELECTOR SWITCH (continued)

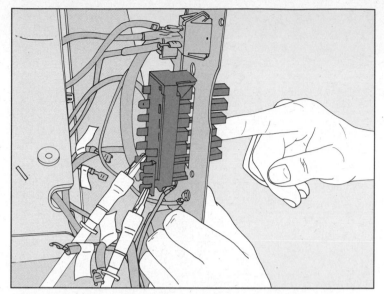

3 **Testing the selector switch.** Set a multitester to the RX1K scale *(page 136)* to measure continuity. Test the pairs of terminals at the switch positions indicated on the chart *(step 2)*. In each case, refer to the chart for the correct reading, usually X for continuity and blank for no continuity. If the switch fails any test, replace it *(step 4)*.

4 **Replacing the switch.** Label and disconnect all selector switch leads, and remove the mounting screws *(above)*. Purchase a replacement selector switch from an air-conditioning parts dealer or from the manufacturer. Install the switch in the air conditioner and connect the leads to the proper terminals. Reassemble and plug in the air conditioner.

TESTING THE COMPRESSOR

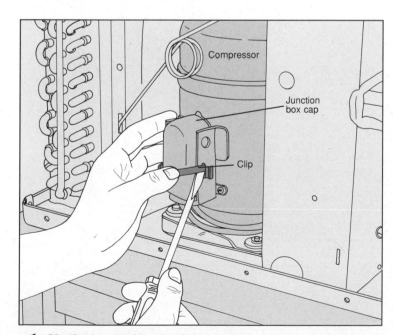

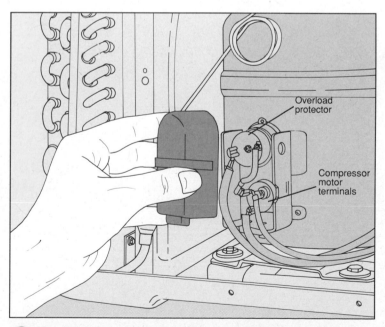

1 **Unclipping the junction box cap.** Unplug the air conditioner and access the internal components *(page 119)*. Discharge the capacitors *(page 125)*. Locate the compressor junction box on the side of the compressor, and use a screwdriver to pry up the clip holding its cap in place *(above)*.

2 **Exposing the compressor motor terminals.** Pull off the junction box cap *(above)* to expose three terminals marked "C", "S" and "R". Label each wire lead for correct reconnection *(page 135)*, then use long-nose pliers to pull the spade lug connectors straight off their terminals. On most air conditioners, the overload protector switch is mounted right beside the compressor motor terminals.

TESTING THE COMPRESSOR (continued)

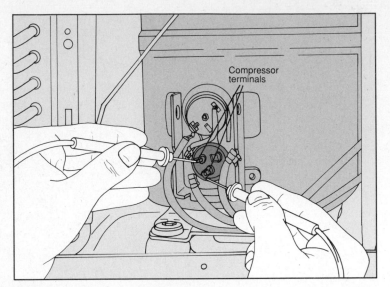

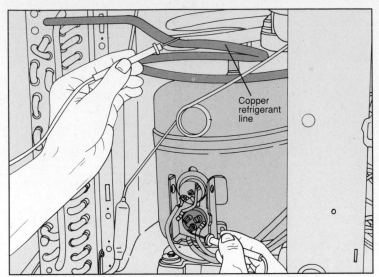

3 **Testing the compressor motor.** Set a multitester to the RX1K scale to test continuity *(page 136)*. Touch one probe to one compressor terminal and one probe to another terminal *(above)*. Test all three possible combinations of terminals. If there is continuity for each test, the compressor motor windings are OK; go to step 4. If there is no continuity in a test, the compressor should be replaced; call for service.

4 **Testing the compressor motor for ground.** Let the compressor cool for three hours. Set a multitester to the RX1K scale to test continuity *(page 136)*. Touch one probe to an unpainted metal surface on the compressor dome or to a copper refrigerant line leading from the compressor, and touch the other probe to each motor terminal, in turn *(above)*. If each test shows no continuity, the compressor is OK. If any test shows continuity, the compressor must be replaced; call for service.

SERVICING THE OVERLOAD PROTECTOR

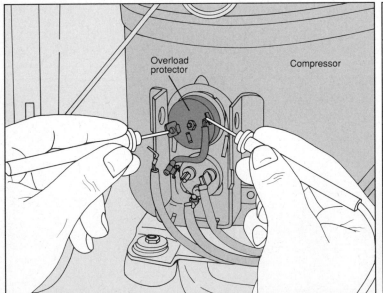

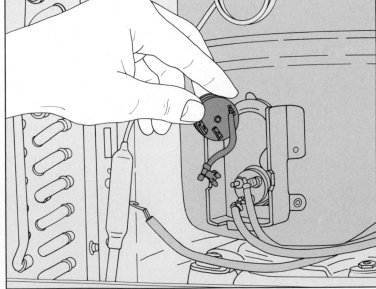

Testing and replacing the overload protector. Unplug the air conditioner and allow the compressor to cool for three hours. Access the internal components *(page 119)*. Remove the junction box cap from the compressor *(page 127)* and locate the overload protector, mounted near the motor terminals. If you do not see the protector, your air conditioner compressor may have an internal one, or none at all; call for service. If your air conditioner has a protector, use long-nose pliers to detach the wire leads from their terminals. Set a multitester to the RX1K scale to test continuity *(page 136)* and touch one probe to each terminal *(above, left)*. If there is continuity, the overload protector is OK. If there is no continuity, unclip the overload protector *(above, right)* and replace it with an identical part, available from a heating and cooling supplies dealer.

TESTING THE FAN MOTOR

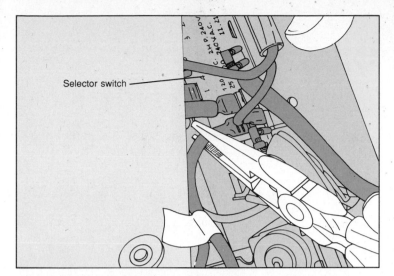

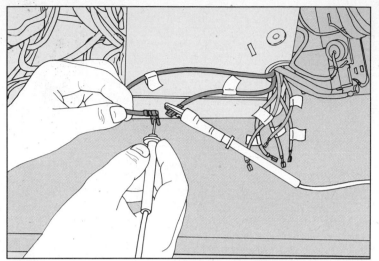

1 **Disconnecting the motor wires.** Unplug the air conditioner and access the internal components *(page 119)*; discharge the capacitors *(page 125)*. Label the motor wires and use long-nose pliers to pull them off the selector switch *(above)*. If your air conditioner has a variable speed fan, there may be eight, nine or more wires connected to its motor.

2 **Testing for continuity.** Set a multitester to the RX1K scale to test continuity *(page 136)*. Locate the common wire, usually the white lead that goes to the selector switch. Touch one probe to the common wire lead and the other probe, in turn, to each of the other wires running from the fan motor to the selector switch. There should be continuity for each test. Next, touch one multitester probe to the motor casing; touch the other probe, in turn, to each of the motor wires. In all cases, there should be no continuity. If the fan motor fails any test, replace it *(below)*.

REPLACING THE FAN MOTOR

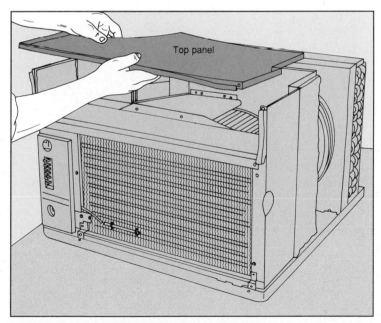

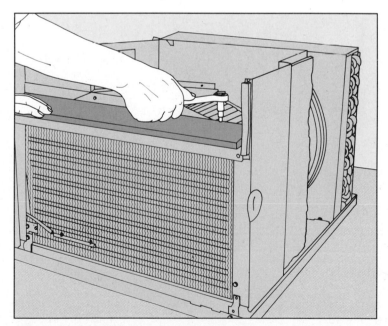

1 **Removing the top panel.** If the air conditioner has a top panel, use a screwdriver or socket wrench to remove the screws or bolts securing the top panel to the chassis, and set the panel aside *(above)*.

2 **Unscrewing the inside panel.** Use a screwdriver or socket wrench to remove any screws or bolts holding a second panel in place over the evaporator coils *(above)*. If your air conditioner is old, the panels may be rusty; replace them with new panels when reassembling the air conditioner.

REPLACING THE FAN MOTOR (continued)

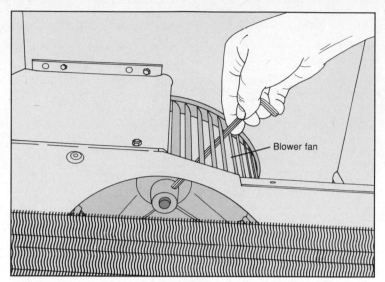

3 **Removing the vent door wire.** Locate the rigid wire connecting the control panel to the vent door. If it runs through a hole in the inside panel, unhook it from the vent door in order to remove the panel from the air conditioner *(above)*.

4 **Loosening the blower setscrew.** Rotate the blower fan by hand to locate the small access hole between two of the blower vanes. Being careful not to bend the vanes, insert a long hex wrench through the hole and loosen the setscrew on the blower fan shaft *(above)*. If you can take out the blower fan at this point, remove it from the motor shaft, then go to step 9. Otherwise, remove the condenser shroud *(step 5)*.

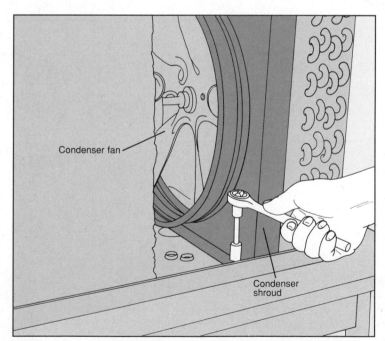

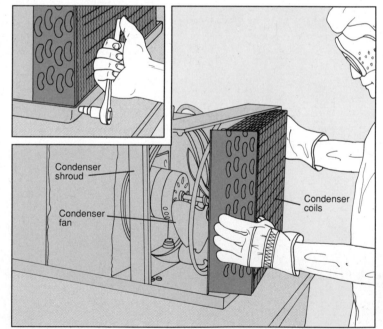

5 **Unscrewing the condenser shroud.** Use a screwdriver or socket wrench to remove the screws or bolts that secure the condenser shroud to the chassis and condenser coils *(above)*. If a retaining brace connects the condenser shroud to the front of the air conditioner, remove it. If you can reach the condenser fan setscrew at this point, loosen the setscrew *(step 7)*, then take out the shroud. Otherwise, first move the condenser coils *(step 6)* to access the motor.

6 **Moving the condenser coils.** With a socket wrench, remove the bolts holding the coils to the base of the air conditioner *(inset)*. Wear safety goggles, heavy gloves and a long-sleeved shirt to protect against high-pressure refrigerant leakage in case the refrigerant line is damaged. Grasp the condenser coils firmly on each side. **Caution:** Carefully ease out the coils only far enough to expose the fan *(above)*.

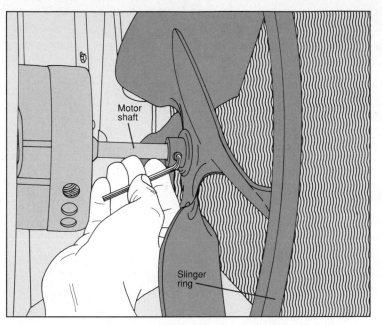

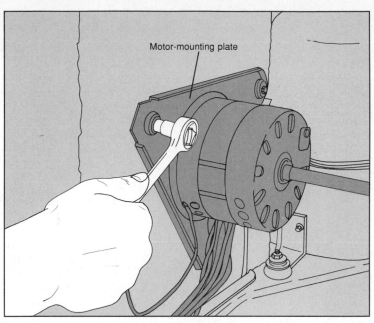

7 **Removing the condenser fan.** Use a hex wrench to loosen the setscrew on the condenser fan shaft *(above)*. Work the fan off the motor shaft, then remove the condenser shroud. Inspect the slinger ring for corrosion; replace it if damaged.

8 **Unbolting the motor.** Label and disconnect all the wire leads *(page 135)* connecting the motor to the control panel, capacitor, compressor and chassis. Next, use a socket wrench to unfasten the bolts holding the motor mounting plate to the dividing wall *(above)*. Do not let the motor fall.

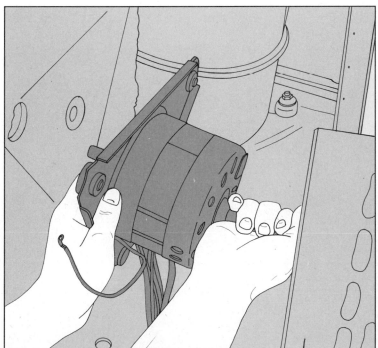

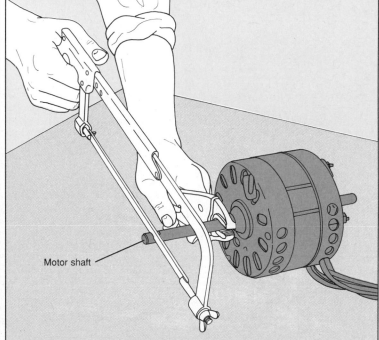

9 **Removing and replacing the motor.** Grasp the motor shaft and the motor mounting plate, and pull the motor out of the air conditioner *(above, left)*. Purchase an identical replacement from the manufacturer, or purchase a universal, or standard, replacement motor from a heating and cooling supplies dealer. Match the motor mount and electrical specifications of the new motor with those stamped on the original motor's housing. If

using a universal replacement, first size the shaft: Measure the length of the old motor shaft and use a hacksaw to cut the new motor shaft to length *(above, right)*. Install the new motor, rewiring it according to the labeled leads on the old motor; follow any other instructions provided with the new motor. Reassemble the air conditioner, reversing the steps taken to disassemble it.

TOOLS & TECHNIQUES

This section introduces basic tools and tests for repairing heating and cooling systems, from stripping and connecting wires to discharging capacitors. You can do many tests and repairs with items already in your toolbox. Specialized tools, such as the inclined manometer and the stack thermometer, can be purchased or rented from a heating and cooling supplies dealer. A capacitor discharger is easy to assemble with a few inexpensive parts available from an electronics supplies store.

A simple battery-powered machine—the multitester—is invaluable for determining whether an electrical part is doing its job. Also called a volt-ohmmeter, it sends a small electrical

current through the part being tested. Two types of multitesters are discussed in this section, analog and digital. Either type can be purchased, or sometimes rented, at an electronics supplies store. Before conducting any tests using a multitester, refer to the instructions in this chapter.

For good repair results, buy the best tools you can afford, use the right tools for the job, and care for them properly. Clean metal tools using a rag moistened with a few drops of light oil (but don't oil the handles). Rub away rust with fine steel wool or emery cloth. Protect tools in a sturdy plastic or metal toolbox, with a secure lock if stored around children.

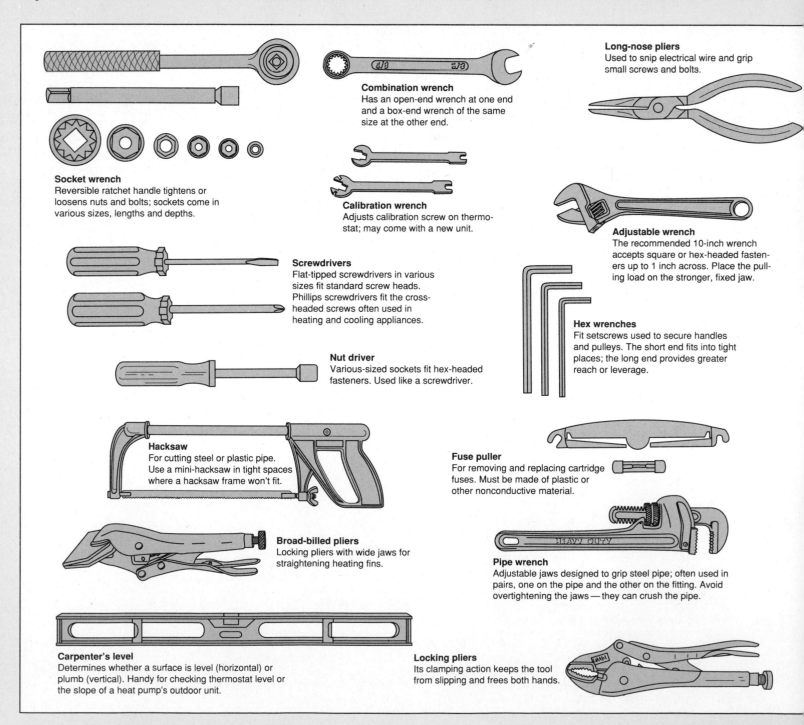

Socket wrench
Reversible ratchet handle tightens or loosens nuts and bolts; sockets come in various sizes, lengths and depths.

Combination wrench
Has an open-end wrench at one end and a box-end wrench of the same size at the other end.

Calibration wrench
Adjusts calibration screw on thermostat; may come with a new unit.

Long-nose pliers
Used to snip electrical wire and grip small screws and bolts.

Adjustable wrench
The recommended 10-inch wrench accepts square or hex-headed fasteners up to 1 inch across. Place the pulling load on the stronger, fixed jaw.

Screwdrivers
Flat-tipped screwdrivers in various sizes fit standard screw heads. Phillips screwdrivers fit the cross-headed screws often used in heating and cooling appliances.

Nut driver
Various-sized sockets fit hex-headed fasteners. Used like a screwdriver.

Hex wrenches
Fit setscrews used to secure handles and pulleys. The short end fits into tight places; the long end provides greater reach or leverage.

Hacksaw
For cutting steel or plastic pipe. Use a mini-hacksaw in tight spaces where a hacksaw frame won't fit.

Fuse puller
For removing and replacing cartridge fuses. Must be made of plastic or other nonconductive material.

Broad-billed pliers
Locking pliers with wide jaws for straightening heating fins.

Pipe wrench
Adjustable jaws designed to grip steel pipe; often used in pairs, one on the pipe and the other on the fitting. Avoid overtightening the jaws—they can crush the pipe.

Carpenter's level
Determines whether a surface is level (horizontal) or plumb (vertical). Handy for checking thermostat level or the slope of a heat pump's outdoor unit.

Locking pliers
Its clamping action keeps the tool from slipping and frees both hands.

Avoid electrical shock by using pliers and screwdrivers with insulated handles, or wrap their handles with electrical tape.

Proper operation of a heating and/or cooling system relies on a flow of electrical current through the unit's circuitry. A key step in troubleshooting a faulty system is simply to inspect the wire connections. Damaged wire ends should be stripped and reconnected as described in this chapter.

A heating and/or cooling system may run on one or more fuel or power sources: electricity, oil or gas. It is important to know where to shut off all its energy supplies before you begin a repair. Turning off electrical power at the service panel and at the unit disconnect switch are described in this chapter. When you are required to close the oil supply valve or gas valve, directions are given in the specific repair.

Heating and cooling repairs can be done safely if you observe basic precautions. Keep the work area well-lit, clean and free of clutter. Use safety equipment such as insulated gloves or safety goggles wherever recommended. Don't smoke or cause a spark around gas or oil burners. When removing sharp access panels, wear work gloves to avoid cuts. Before working on an air conditioning unit or heat pump, prevent electrical shock by discharging the capacitors when recommended.

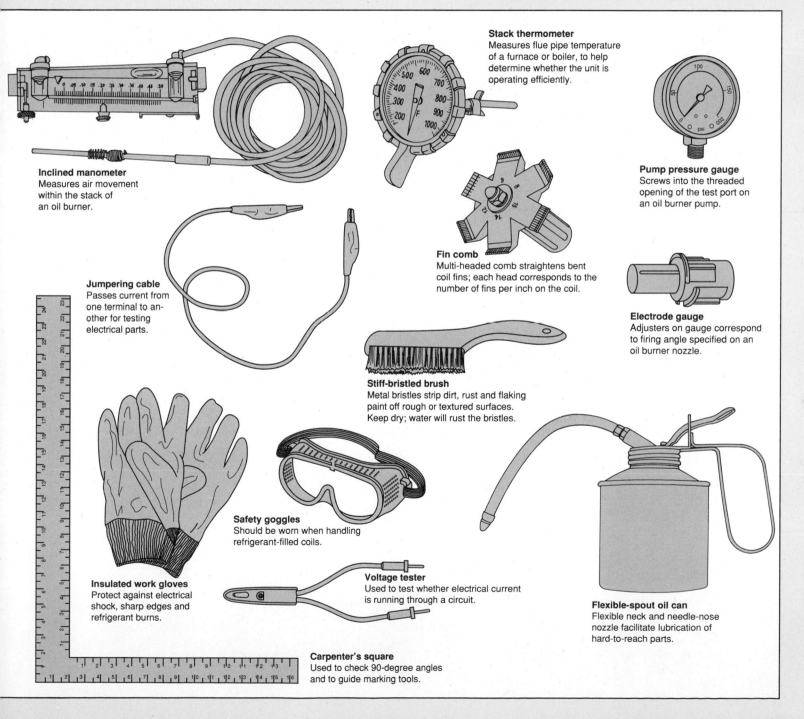

Inclined manometer
Measures air movement within the stack of an oil burner.

Stack thermometer
Measures flue pipe temperature of a furnace or boiler, to help determine whether the unit is operating efficiently.

Pump pressure gauge
Screws into the threaded opening of the test port on an oil burner pump.

Fin comb
Multi-headed comb straightens bent coil fins; each head corresponds to the number of fins per inch on the coil.

Jumpering cable
Passes current from one terminal to another for testing electrical parts.

Electrode gauge
Adjusters on gauge correspond to firing angle specified on an oil burner nozzle.

Stiff-bristled brush
Metal bristles strip dirt, rust and flaking paint off rough or textured surfaces. Keep dry; water will rust the bristles.

Safety goggles
Should be worn when handling refrigerant-filled coils.

Insulated work gloves
Protect against electrical shock, sharp edges and refrigerant burns.

Voltage tester
Used to test whether electrical current is running through a circuit.

Flexible-spout oil can
Flexible neck and needle-nose nozzle facilitate lubrication of hard-to-reach parts.

Carpenter's square
Used to check 90-degree angles and to guide marking tools.

TURNING OFF POWER AND SERVICING BREAKERS AND FUSES

For most repairs and inspections in this book, the first — and most important — step is to turn off power to the heating or cooling unit. Simply switch off a window air conditioner, then unplug it from the electrical outlet. For other heating and cooling units, turn off power at the main service panel *(below)* and at the unit disconnect switch *(page 135)*, if your unit has one. As an extra precaution, after you have turned off power to a baseboard heater or line-voltage thermostat, check to make sure that the power is indeed off. Conduct the test shown for baseboard heaters on page 85, and the test for line-voltage thermostats on page 22. Use a similar precaution when testing voltage at the transformer. Never touch tester probes, wires or any part of the unit while the power is on.

If the unit does not run at all, a circuit breaker may be tripped, or one or two fuses blown. Reset breakers and replace fuses as described below.

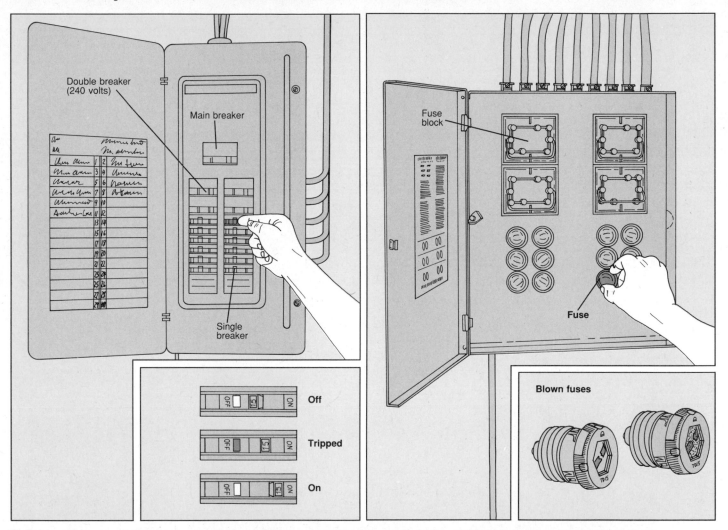

Turning off power at the service panel. If your home was built or renovated after 1960, the electrical system is probably protected against overload by a circuit breaker panel *(above, left)*. Older systems have fuse panels *(above, right)*. In many cases, a single breaker or fuse serves a heating or cooling unit; but most central air conditioning units are linked to a double breaker, and heat pumps use one or two double breakers. Look on the service panel — usually below the main breaker or fuse blocks — to identify the unit's breaker or fuse. If this information is not written on the service panel, turn off each breaker or remove each fuse until power to the system or unit goes off; take this opportunity to label the service panel for future reference. Move a breaker all the way to the OFF position *(above, left)*; it may spring back to an intermediate position. On a fuse panel, unscrew the fuse by holding its outer edges and twisting it counterclockwise *(above, right)*. Before working on the unit, also turn off power at the unit disconnect switch *(page 135)*. After completing a repair, restore power at the service panel, then at the unit disconnect switch.

Resetting breakers and replacing fuses. A tripped circuit breaker *(above left, inset)* will be in an intermediate position between ON and OFF. To reset the breaker, push it to OFF, then to ON. A blown fuse *(above right, inset)* has a broken or melted filament, or discolored glass. Unscrew the fuse, and screw in a replacement of exactly the same rating.

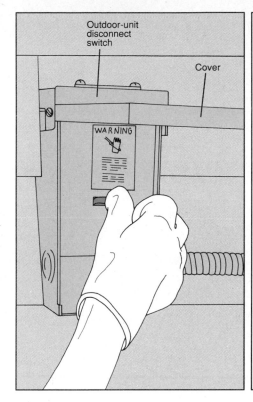

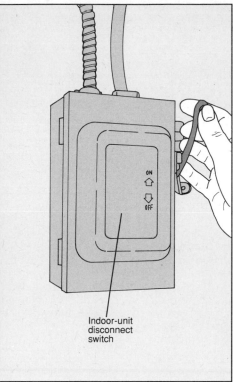

Turning off power at the unit disconnect switch. Locate the unit disconnect switch. It may be indoors, near the furnace or boiler, or outdoors, mounted on or near the heat pump or central air conditioning unit. If the floor or ground in that vicinity is wet, stand on a dry board or rubber mat, or wear rubber boots; also wear heavy rubber gloves. Using one hand, shift an indoor-unit disconnect switch to the OFF position *(near left)*. On an outdoor-unit disconnect switch, lift up the weatherproof cover *(far left)* to access the ON/OFF switch. As an extra safety precaution, also turn off power to the heating or cooling unit at the main service panel *(page 134)*. After completing a repair, restore the power at the service panel, then at the unit disconnect switch.

IDENTIFYING WIRES FOR CORRECT RECONNECTION

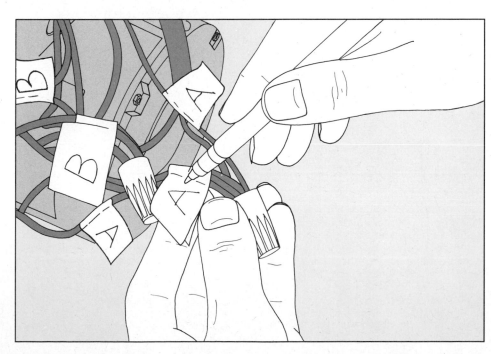

Labeling wires. Before disconnecting wires from screw terminals, spade lug terminals or wire caps, label them to ensure correct reconnection. In the case of especially complicated wiring, also draw a diagram of the wire and terminal locations. Wrap each wire with a 2-inch piece of masking tape, then use a pen to mark each tag with the wire's terminal location. For bundles of wires joined by wire caps, identify each wire as belonging to bundle A or bundle B *(left)* before removing the wire caps.

TROUBLESHOOTING WITH A MULTITESTER

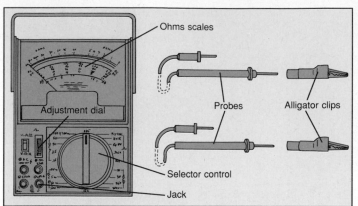

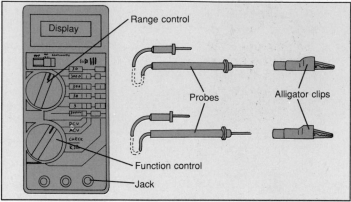

Using a multitester. A multitester tests continuity *(below)* and resistance *(page 137)* by sending a low-voltage electrical current from its batteries through the part being tested. It displays, in ohms, the precise amount of resistance the part has to electrical current. Zero ohms—total lack of resistance—indicates a completed circuit, or continuity. A multitester is also used to test voltage *(page 137)*, measuring the precise amount of electrical current flowing in a completed circuit. A multitester may be analog *(above, left)*, with a needle that sweeps across a scale, or digital *(above, right)*, with a numerical display. On either type,

connect the black cable to the negative jack and the red cable to the positive jack. To use an analog multitester, first "zero" the meter: With the multitester set at RX1, touch the probes together, or clip the alligator clips together. The needle should sweep from left to right toward ZERO; turn the adjustment dial until the needle lies directly over ZERO. (If the multitester won't "zero" the batteries are low.) A digital multitester can be easier to use and read; consult the instructions that come with it. When testing, the alligator clips or probes should firmly contact bare metal, terminals or wire ends; not insulated, painted or dirty ones.

TESTING CONTINUITY

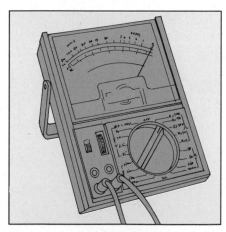

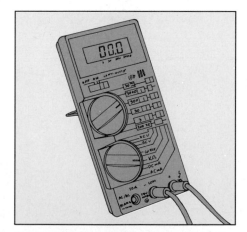

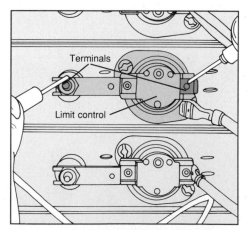

Using an analog multitester. Prepare the multitester as described above. Set the selector control to the RX1K setting *(above)*, or other setting specified for the repair. Touch each probe — or attach each clip — to a terminal of the component you are testing, as shown at far right. If the needle swings to zero ohms, there is continuity. If the needle does not move from infinite ohms, there is no continuity.

Using a digital multitester. Prepare the multitester as described above. With the multitester turned on, turn the range control to its lowest setting and set the function control to OHMS *(above)*. Touch each probe — or attach each clip — to a terminal of the component you are testing, as shown at right. If the reading is 00.0, or if you hear a beep, there is continuity.

Testing continuity. A limit control in an electric furnace is being tested for continuity *(above)*. When each probe contacts a switch terminal the multitester tries to send low-voltage electrical current from its batteries through one probe. If the current passes through the control to the other probe, there is a complete circuit and continuity. If the current does not pass through to the other probe, there is no complete circuit and no continuity.

TESTING VOLTAGE

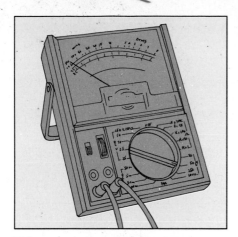

Using an analog multitester. Prepare the multitester as described on page 136. Set the selector control to 50 ACV for a transformer, or to the lowest setting on the DCV scale for a thermocouple. Turn off power to the system *(page 134)*. Attach each alligator clip to a terminal of the component, as shown at far right. Turn on the power; the multitester needle will show the amount of voltage in the circuit.

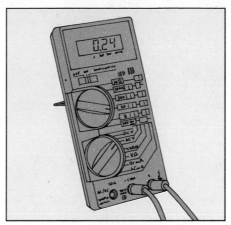

Using a digital multitester. Prepare the multitester as described on page 136, and turn it on. Set the range control to 30. If testing a transformer, set the function control to ACV. Turn off power to the system *(page 134)*. Attach each alligator clip to a terminal of the component, as shown at right. Turn on the power; the multitester display will show the amount of voltage in the circuit.

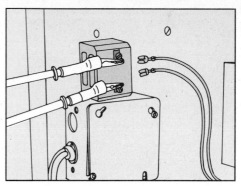

Testing voltage. Caution: Make sure power is off when attaching multitester clips to the component you are testing; do not touch the clips or the unit while power is on. In the example above, the power from a gas furnace transformer is tested. With each clip contacting a transformer terminal *(above)*, a low-voltage electrical current is sent through the multitester by turning on power to the furnace. The amount of electrical current flowing through the component is shown by the multitester's needle or display.

TESTING A CAPACITOR

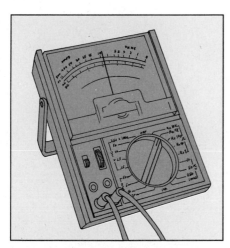

Using an analog mutitester. Prepare the multitester as described on page 136. Set the selector control to RX1K. Touch each probe—or attach each clip—to a terminal of the component you are testing, as shown at far right. The needle position on the scale indicates the resistance level.

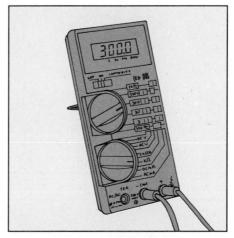

Using a digital multitester. Prepare the multitester as described on page 136. With the multitester turned on, set the range control to its lowest setting and set the function control to OHMS. Touch each probe—or attach each clip—to a terminal of the component you are testing, as shown at right.

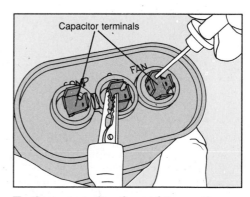

Capacitor terminals

Testing a capacitor for resistance. On a capacitor with two or three *(above)* terminals (with or without a bleed resistor), touch the red multitester probe to the positive (+) capacitor terminal, and the black probe to the common, or negative (-), terminal. As the multitester tries to send low-voltage electrical current from its batteries through the capacitor, resistance increases within the capacitor. The needle should swing toward zero ohms (no resistance), then back toward infinity.

WORKING WITH WIRE

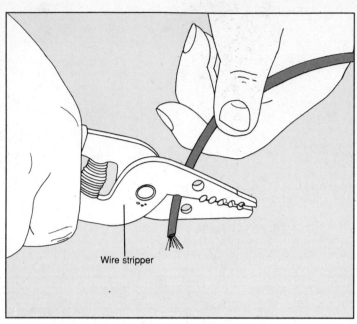

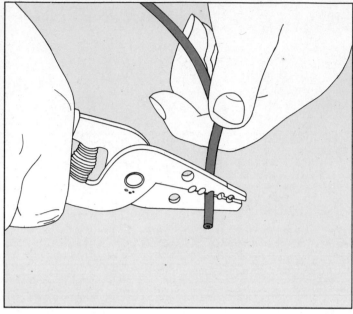

Wire stripper

Stripping wire insulation. A wire stripper removes insulation cleanly without damaging the wire inside. Graduated notches accommodate the standard wire gauges, and it has a wire cutter near the joint. When a wire end at an electrical connection is frayed or damaged, use the wire stripper's cutting edge to snip off the damaged end *(above, left)*. Then insert 3/4 inch of the wire into a matching notch on the stripper. Close the tool and twist it back and forth gently until the insulation is severed *(above, right)*, then pull it off the wire.

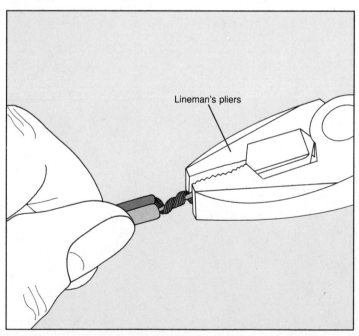

Lineman's pliers

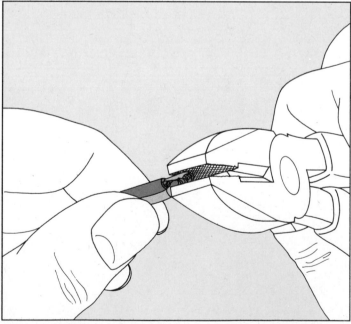

Joining stranded wire to solid wire. Holding the two wire ends parallel, wrap the stranded wire in a clockwise spiral around the solid wire *(above, left)*. With sturdy pliers such as lineman's pliers, fold the end of the solid wire over the wrapped portion *(above, right)*. Screw on a wire cap *(page 139)* to secure the connection. To join stranded wire to stranded wire, hold the wire ends parallel and twist them together in a clockwise direction, then install a wire cap.

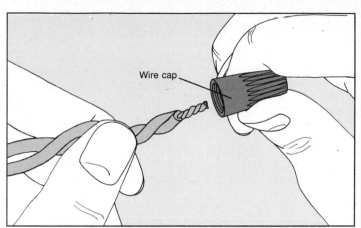

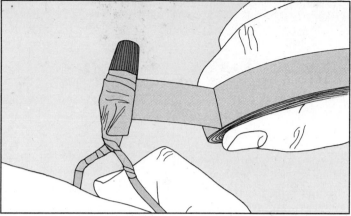

Installing a wire cap. Using a wire stripper, remove an equal length of insulation from each wire end, and twist the wires together *(page 138)*. Slip a wire cap over the connection *(above, left)*. Screw the cap clockwise until it is tight and no bare wires remain exposed. Test the connection with a slight tug on the wire cap. To secure the connection, wrap electrical tape around the base of the cap, then once or twice around the wires, and finally around the base of the cap again *(above, right)*.

INSTALLING CRIMP CONNECTORS

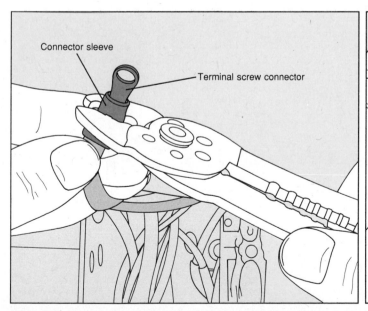

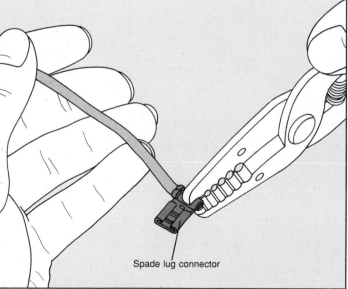

Crimping a connector. Purchase UL approved crimp connectors — also called solderless connectors — at an electrical parts supplier. Buy the right size and style to fit the terminals and the wire gauge. To install a crimp connector, strip back the wire *(page 138)*, fit it into the connector sleeve and crimp it using the outer notch on a wire stripper *(above)*.

IDENTIFYING CAPACITORS

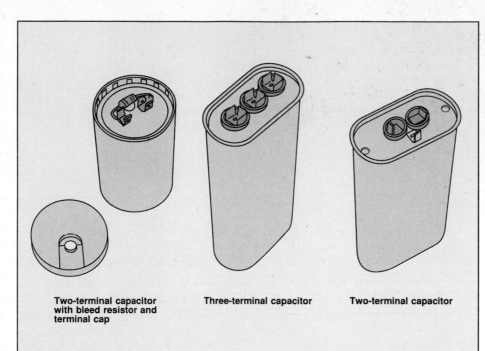

Three types of capacitors. The capacitors in heating and cooling units store electricity, which is used to start a motor or help the motor run more efficiently. A capacitor may have two or three terminals *(left)*. On some two-terminal capacitors, a bleed resistor soldered across the capacitor's terminals discharges stored electricity when the motor is off. Three-terminal capacitors are actually two capacitors housed in one casing. They are used to help start and run motors. This type of capacitor has a common terminal located between the other two terminals. If you must work close to capacitors while servicing heating or cooling equipment, always discharge the capacitors first *(below)*.

Two-terminal capacitor with bleed resistor and terminal cap

Three-terminal capacitor

Two-terminal capacitor

DISCHARGING CAPACITORS

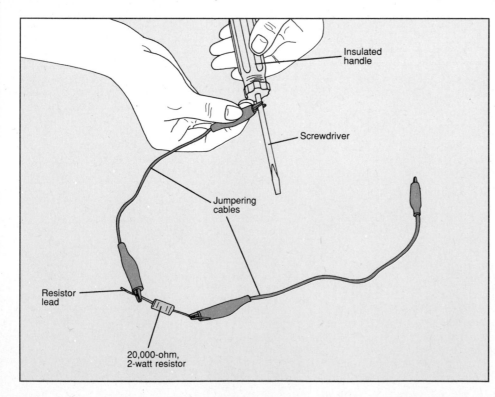

Insulated handle

Screwdriver

Jumpering cables

Resistor lead

20,000-ohm, 2-watt resistor

Making a capacitor discharging tool. This simple tool discharges capacitors without damage to them or injury to you. Assemble two jumpering cables with alligator clips; a 20,000-ohm, 2-watt resistor, available from an electronics supplies store; and a screwdriver with an insulated handle. Clip one end of a jumpering cable to one resistor lead and clip the other end to the blade of the screwdriver. Clip one end of the other jumpering cable to the remaining resistor lead *(left)*.

DISCHARGING CAPACITORS (continued)

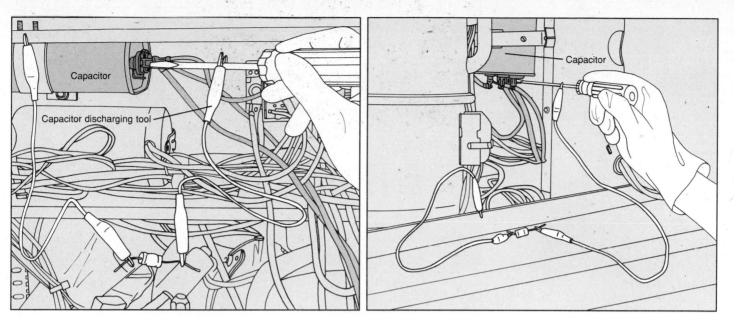

Discharging a capacitor with exposed terminals. Turn off power to the heating or cooling unit *(page 134)*. Using a capacitor discharging tool *(page 140)*, attach the free alligator clip to a clean, unpainted metal part of the unit's chassis. Hold the insulated handle of the screwdriver with one hand and touch the screwdriver blade to each of the capacitor's two *(above, left)* or three *(above, right)* terminals, in turn, for one second. Discharge all capacitors in the unit.

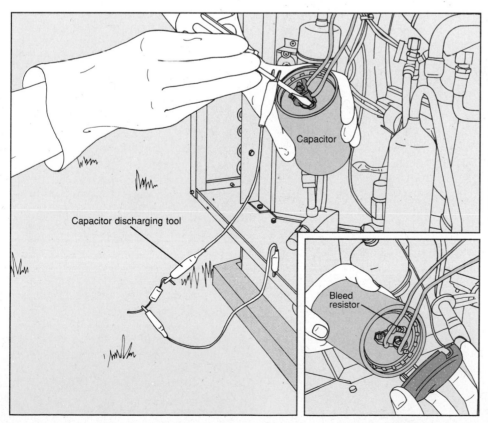

Discharging a capped capacitor. Turn off power to the unit *(page 134)*. Attach the free alligator clip of a capacitor discharging tool *(page 140)* to a clean, unpainted metal part of the unit's chassis. Release the capacitor from its clips and carefully remove the cap covering the capacitor terminals *(inset)*. **Caution:** Do not touch the capacitor terminals. Hold the capacitor in one hand and, holding the insulated handle of the screwdriver with the other hand, touch the screwdriver blade to each terminal on the capacitor, in turn, for one second *(above)*. Discharge all capacitors in the unit.

INDEX

Page references in *italics* indicate an illustration of the subject mentioned.
Page references in **bold** indicate a Troubleshooting Guide for the subject mentioned.

ACKNOWLEDGMENTS

The editors wish to thank the following:
Alberta Department of Energy, Energy Conservation Branch, Edmonton, Alta.; American Society of Heating, Refrigerating and Air Conditioning Engineers, Atlanta, Ga.; Kevin Barrett, American Air Filter, Louisville, Ky.; Rolland Bertin and Clifford Taite, Brock Engineering Mfg. Co. Ltd., Montreal, Que.; Boston Edison Co., Boston, Mass.; Roy Bradley, Canadian Plumbing and Heating Supplies Ltd., Montreal, Que.; Canadian Gas Association, Don Mills, Ont.; Canadian Institute of Plumbing and Heating (Canadian Hydronics Council), Toronto, Ont.; Cos Caronna, Lavergne, Tenn.; Serge Charbonneau, Gaz Metropolitain, Montreal, Que.; David J. Chase, Dallas, Tex.; Richard Day, Palomar Mountain, Calif.; Paul de Wit, Honeywell Canada Ltd., Montreal, Que.; Ted Dreyer, Residential Division, Honeywell Inc., Golden Valley, Minn.; Dwyer Instruments Inc., Michigan City, Ind.; Energy, Mines and Resources Canada, Montreal, Que.; Melanie Gagnon, Montreal, Que.; Robert E. Hensley, Montreal, Que.; Hydro-Quebec, St. Laurent, Que.; Fred Kokin, Montreal, Que.; Michael A. MacDonald, Montreal, Que.; Ontario Hydro, Toronto, Ont.; Jean-Pierre Robbe and Martin Saltzman, S. Albert & Co. Ltd., Montreal, Que.; Steven Sacks, Les Services Organon, Montreal, Que.; Dr. M. Shulman, Montreal, Que.; Whirlpool Corporation, Benton Harbor, Mich.

The following persons also assisted in the preparation of this book:
Arlene Case, Richard Fournier, Patrick J. Gordon, Renaud Kasma, Nancy D. Kingsbury, Julie Léger, Caroline Miller, Natalie Watanabe and Billy Wisse.

Typeset on Texet Live Image Publishing System.